AF820418

Copyright © 1988
by
Marcus Books
Post Office Box 327
Queensville, Ontario
Canada L0G 1R0

(Marcus Books is a trading style
of 703809 Ontario Ltd.)

Write for free catalogue of books,
audio and video tapes, and other items.

All rights reserved.
No part of this book may be reproduced in any
form without the written permission of the publishers.
ISBN 0-919951-42-2

First printing October 1988
Manufactured in Canada by Webcom Limited

Cover layout and design by Paul Chaput

Cover art work by

THE MARS CONNECTION

by
Maurice B. Cooke

CONTENTS

Section 1: The Research

Section 2: Mars Conjunctions

Section 1
The Research

INTRODUCTION

It seems somehow appropriate to be putting together a book on Mars just as the red planet is swinging around to its closest "close" approach in several decades (summer of 1988). Although this body does pass every couple of years through its "perigee", or nearest approach to the earth, the actual minimum distance itself varies on a much longer cycle. Perhaps this astronomical event will serve to arouse some curiosity among those who would not otherwise think of investigating unconventional influences on the collective psyche. Regardless of how the reader has been led to pick up this book, it is appropriate here to offer a few preliminary remarks.

This book has several goals:

Firstly, it summarizes research which I carried out in the late 1960's, and which led to publication of the book, *Astrological Warnings and the Stock Market* (1970, Pagurian Press, Toronto). This work was reprinted in 1979, in an updated version called *Sun Spots, Stars and the Stock Market*. The latter contained a clear prediction that the market would touch a low point in late 1987. I had forgotten that the book had mentioned such a likelihood, until a friend pointed out that this prophecy appeared to have correctly foretold the "crash" of October 19, 1987. Since the earlier books were now out of print with no possibility of reprinting, I decided to put together a brief summary of the reasoning and the findings contained in the earlier works, and to offer a preview of possible market movements through the year 2000. The first three chapters of this book deal with the Astrology of the stock market.

Secondly, the present book reprints the gist of an article

which I wrote some time ago summarizing the results of research into the Astrology of Alcoholism.

Thirdly, it contains a study of the influence of the planet Mars on outbreaks of armed conflict.

The topics mentioned above should be of interest to brokers, investors and historians, as well as to students and practitioners of Astrology. In order to make the book accessible to everyone regardless of previous exposure to the subject, I have explained all terms and assumptions carefully.

The remainder of this book consists primarily of material that should prove helpful in the analysis of individual birth charts. It was prepared in the same manner as the majority of the books in the *Hilarion Series,* and should be of interest not only to Astrologers, but to all those who seek the "reality behind reality".

Maurice B. Cooke
Toronto, Canada.
August, 1988

Chapter 1

Mars and the Dow-Jones

In 1970, I wrote a small book entitled *Astrological Warnings and the Stock Market,* which was published by Pagurian Press (Toronto) in hardback under the name T. Rieder. The book was reprinted in 1975, and then was revised and updated for a new edition appearing in 1979 under the title, *Sun Spots, Stars and the Stock Market*. Both are now out of print.

One of the most fascinating correlations explained in that book relates to the clear influence of the planet Mars on human economic behavior, as reflected in the stock market averages. My studies leading to the preparation of *Astrological Warnings and the Stock Market* involved, among other things, tracing all conjunctions of Mars with the so-called "Heavy" planets (Jupiter, Saturn, Uranus and Neptune) over a 70-year period, and noting the behavior of the Dow Jones Industrial Average (DJIA) before and after the point of each conjunction. Pluto was omitted, because to introduce another category of Martian conjunction would have produced too many time points with possibly overlapping effect. In what follows, I will refer to the planets Jupiter, Saturn, Uranus and Neptune as the "heavy planets".

I soon concluded that a conjunction between Mars and one of the heavy planets does not by itself produce any clear reaction in the DJIA. However, I noted that whenever a conjunction between Mars and one of the four heavies is simultaneously afflicted by a square or opposition from another one of the heavies within an appropriate orb, the DJIA tends to react negatively about 70% of the time.

For those not familiar with astrological terminology, it is appropriate here to explain some of the basics of the subject.

The sun, moon and all of the planets in the solar system appear to trace their paths along or close to an imaginary line through the fixed stars. That line is called the Ecliptic, and represents the path of the sun itself. The various planets and the Lights (sun and moon) all move at different speeds with respect to the fixed backdrop of stars. The sun of course completes one circuit in one of our years. The moon takes about 27 days. Mars finishes its journey in roughly 2 years, Jupiter in about 12 years, Saturn in 29 years, Uranus in 84 years and Neptune in 165 years.

Due to these different speeds, the various planets appear to conjoin each other at different times. Mars, moving at a relatively lively pace, conjuncts each of Saturn, Uranus and Neptune roughly every 2.3 years, and laps Jupiter on a slightly longer cycle.

I mentioned earlier my discovery that a conjunction of Mars with one of the heavy planets, where the point of conjunction is simultaneously afflicted by a square or opposition from one of the other heavy planets, tends to correlate with a negative market reaction. I defined "negative reaction" to include the following:

a) DJIA dropping both before and after the time of the conjunction;

b) DJIA stable (flat) before and dropping after the time of the conjunction;

c) DJIA rising before and dropping after the time of the conjunction;

d) DJIA rising before and stable (flat) after the time of the conjunction;

The reference to squares and oppositions deserves some elucidation, for the sake of those readers not fully conversant with astrological terminology. Because the planets all appear to move through the fixed stars at different speeds, it will be obvious that they will, from time to time, form different angles as observed from the earth. If one looks in a given direction to find the planet Mars for example, and then turns through exactly 90 degrees (a right angle) to find, say, the planet Saturn, then in astrological parlance these two planets are said to be "squaring" each other, or "in square". If one must look in diametrically opposite directions to find two given planets (as we must do to find the sun and the moon at the time of full moon, for example), then the bodies are said to be "in opposition". Generally speaking, the square and the opposition are considered to be "difficult" aspects (aspects means angles), wherein the energies of the planets do not blend easily. The influence of each planet in the square or opposition is considered to be made more negative or difficult due to the effect of the aspect from the other planet. By the same token, the angle 60 degrees (called "sextile") and the angle 120 degrees (called "trine") are looked on generally as favorable or positive influences on the planets concerned, tending to bring forth the best side of each planet's realm of influence.*

Several instructive examples of the effect under discussion can be seen by looking at Figure 1, which is a chart of the Dow-Jones Industrial Average for the years 1907-1912, plotted using month-end figures.

The first point of interest is on July 17, 1907, when Mars conjoined Uranus at roughly 10 degrees of Capricorn, while simultaneously opposing Neptune within 3 degrees. In the 5 months following the conjunction, the DJIA declined by 35%.

On December 28, 1909, Mars was conjunct Saturn at about 16 degrees of Aries, while simultaneously square (90 degrees) to Neptune within 2 degrees. The DJIA, which had risen strongly in the 1908-1909 bull market, suddenly turned around and over the next 7 months backed off by 22%.

On August 16, 1911, the Mars-Saturn conjunction at 20 degrees of Taurus was widely opposed by Jupiter at about 8

---

* Note that Hilarion, in his book *Astrology Plus,* modifies this scheme somewhat by stating that Saturn virtually *always* brings out a negative side of another planet to which it forms an aspect, regardless of whether the aspect is traditionally considered positive or negative. He also holds that Jupiter virtually *always* throws a benefic influence on any other planet with which it is in aspect, again regardless of the specific angle.

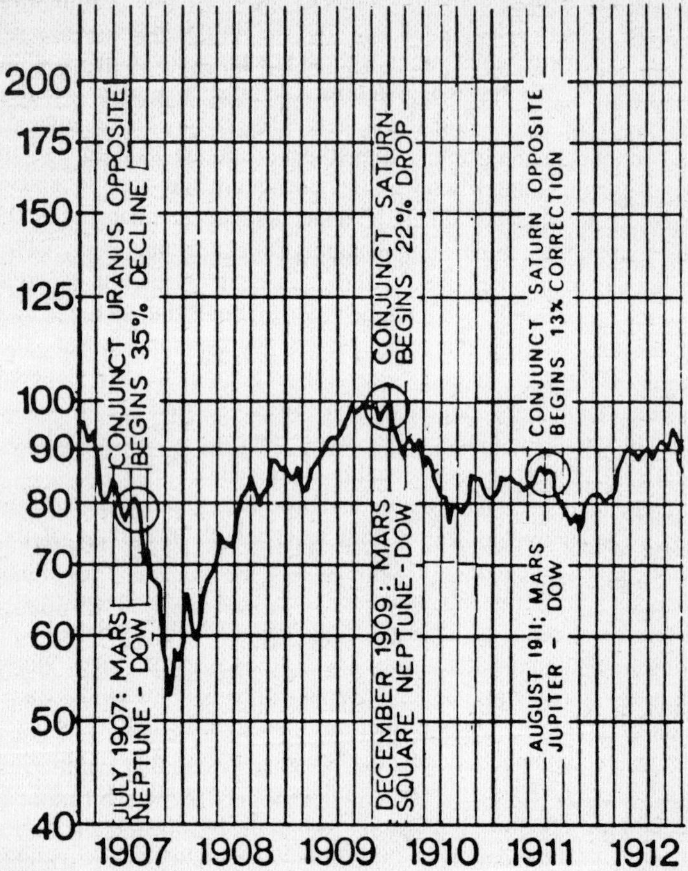

Figure 1

degrees of Scorpio. The Dow-Jones at that point began a minor correction of 13%.

Now, in order to comprehend the results of the various tallies which I summarized in the earlier books, it is necessary to explain the concept of orbs. In astrological parlance, the word "orb" designates the amount by which a given aspect between two planets is not exact. For example, imagine that Mars were at exactly 4 degrees of Aries, while Saturn were at exactly 7 degrees of Aries. The "orb" of this conjunction would be 3 degrees (i.e. the arc by which the planets are separated). This example is simplified by the fact that we are assuming whole degrees. However, each degree is divided into 60 "minutes of arc" (not to be confused with minutes of time), and therefore it is possible to specify planetary positions with greater precision than to a whole degree.

Consider another example. If Jupiter were at 13 degrees, 23 minutes of Virgo, while Uranus were at 10 degrees, 11 minutes of the same sign, the conjunction would have an orb of 3 degrees, 12 minutes. For convenience we shall write this as 3 deg 11. Now all aspects can be looked on in the same way. For example a square or an opposition can also be defined as having a specific orb. Consider the signs Cancer and Capricorn, which are exactly opposed to each other around the zodiac (ecliptic). This means that the location 1 Cancer is exactly opposite the location 1 Capricorn. Now if two planets were located at exactly 1 degree of these two signs, they would be in exact opposition, and the orb would be zero. However, if they were at 5 deg 45 Cancer and 7 deg 56 Capricorn respectively, the opposition would have an orb of 2 deg 11.

As a general rule, oppositions and conjunctions are considered to be operative at any orb up to about 8 degrees, while the semi-sextile (30 deg), sextile (60 deg), square (90 deg), trine (120 deg) and inconjunct (150 deg) are all permitted an orb of about 5 degrees. (Some astrologers allow a bit more in each case.) There is general agreement that the comparative influence or prominence of a given planetary pattern is roughly inversely proportional to the size of the orb. Thus, the impact of a given pattern in one's birth-chart varies from very prominent at an orb of 2 degrees or less, to rather weak at the outside orb limit permitted for the particular aspect.

Returning now to the stock market, it was necessary for me to establish fairly rigorous definitions for the various planetary patterns, prior to determining whether there were any significant correlations with the movements of the Dow-Jones. Accordingly, for each Martian pattern, I allowed a count of −1 for the conjunction itself, a count of −1 for a square within an orb of 5 deg 20, a count of −½ for a square whose orb was between 5 deg 20 and 8 deg 0, a count of −1 for an opposition within an orb of 8 deg 0, and a count of −¾ for an opposition whose orb was between 8 deg 0 and 11 deg 0. Outside of these orb limits, no value was assigned.

Thus in the examples given earlier, the pattern of December 28/09 had a value of −2 (−1 for the Martian conjunction plus −1 for the square to Neptune within 5 deg 20), while the arrangement on Aug 16/11 had a value of only −1 (the conjunction), because the opposition was too wide to count.

I had carefully plotted the DJIA from 1899 to 1971 inclusive, and I was ready to make the tally. A list was prepared, containing all Martian patterns with a count greater than −1½, and opposite each entry I placed a line sketch identifying what the Dow-Jones had done immediately before and immediately after the time-point of the Martian conjunction. The list contained 55 entries and appeared as Appendix B in the original book.

Previously in this chapter I have identified the Dow-Jones patterns that I considered to be negative.

Three counts were then made, as summarized below:

1st count: take all entries in the list.

Of 55 total,	26 tallied with a bad mkt. pattern	= 47%
	10 tallied with a neutral mkt. pattern	= 18%
	19 tallied with a good mkt. pattern	= 35%

2nd count: take only those entries in the list which have a value of −2 or worse.

Of 40 total,	24 tallied with a bad mkt. pattern	= 60%
	7 tallied with a neutral mkt. pattern	= 17.5%
	9 tallied with a good mkt. pattern	= 22.5%

3rd count: take only those in the second count which were either unalleviated by sextiles or trines or, if alleviated, were such that the closest aspect was an afflicting aspect (square or opposition) rather than an alleviating aspect (trine or sextile).*

* Because the trines and sextiles are considered "positive" aspects, it is legitimate to regard their presence in an otherwise negative pattern as tending to lift or alleviate the 'heaviness' of the pattern.

Of 33 total,	23 tallied with a bad mkt. pattern	= 70%
	4 tallied with a neutral mkt. pattern	= 12%
	6 tallied with a good mkt. pattern	= 18%

Using the tools of statistics, it is possible to calculate the "significance level" of the result of the third count. Firstly, by taking a large number of randomly selected dates during the period under examination, and assessing the market pattern around those random dates, one finds that "bad" patterns occur only about 37% of the time. Statistical calculation reveals that the odds are more than 10,000 to 1 against finding that 70% of any given set of dates will correlate with "bad" market patterns, given that random dates yield only 37%.

A further count was carried out in the basic list of 55 Martian patterns to learn the frequency with which individual "heavy" planets took part in the patterns that correlated with bad markets. Of the 13 instances where Mars conjoined Saturn, 9 corresponded with bad market patterns, i.e. about 69%. Also, of the 5 instances where Mars opposed Jupiter, none corresponded to a "good" market pattern (i.e. all were bad or neutral). Of the ten patterns where Mars squared Neptune, seven coincided with a bad market picture.

From the foregoing investigations, I drew the following general rules:

1.(a) Mars appears to function as a "trigger" for "bad" or undesirable long-term stock market movements, whenever the red planet is found in certain geometric patterns with two or more of the planets Jupiter, Saturn, Uranus and Neptune.

(b) Historically, whenever Mars conjoins one of the heavy planets named above under (a) and is simultaneously in square or opposition to another of these planets, there is a distinct tendency for the Dow-Jones to show a bad or negative pattern, so long as the inexactitude of the square or opposition is sufficiently small.

(c) The effective orb for a square is in the region of 5 deg 20, while that for an opposition is in the neighborhood of 8 degrees.

(d) The greater the number of planets involved in a given negative pattern (within the indicated orbs), the greater is the likelihood for a "bad" market pattern to occur.

2. The Dow-Jones is about 70% likely to perform a negative pattern whenever Mars conjoins one of the planets named in 1(a) while being simultaneously in square or opposition aspect to another one of the same planets, within the orbs set out in 1(c), provided the negative planetary geometry is not alleviated by trines or sextiles, or if alleviated, the closest aspect is a square or opposition.

3. The geometries defined above under point 2 are even more inauspicious if:

(a) Mars conjoins Saturn
(b) Mars squares Neptune
(c) Mars opposes Jupiter.

4. When a planetary pattern is alleviated by a trine or a sextile, and the closest aspect is one of the latter rather than a square or opposition, the percentage likelihood of the market showing a negative pattern must be decreased.

From additional investigations and tallies detailed in the earlier books but not discussed in this chapter, I drew the following further conclusions:

5. The basic Martian patterns are more likely to coincide with a "bad" market if the secondary aspect to Mars is an opposition, regardless of the planet.

6. All Martian retrograde stations and all major planet afflicted patterns (i.e. not including the planet Mars) must be regarded as having the potential for negatively influencing the market, even though it is not possible to assign to them a specific probability that they will coincide with a negative market pattern. Hence, where the market has strongly advanced over a period of months or years, a forthcoming Martian retrograde station or a pattern in which two of the heavy planets are conjunct while simultaneously squaring or opposing another one of the heavy planets can be regarded as a possible turnaround point.

The reader is doubtless wondering whether I discovered any patterns that coincided with upturns in the Dow-Jones, but here he will be disappointed. Such patterns may well exist, but I could not unearth any that boasted a respectable significance level when submitted to the scrutiny of statistical analysis. The best advice I can offer to the investor is to time his purchases at the start of a longish period which is free of unalleviated negative planetary patterns as defined above.

There is, however, one further important factor that should not be overlooked, and this relates to the uncanny cycle of 4½ years which the Dow-Jones has meticulously followed ever since the end of the Second World War. I say uncanny because the cycle corresponds almost exactly with twice the Martian synodic period. It is as if the market has fallen under the sway of the red planet, and is powerless to halt the relentless cosmic dance that ties it so closely with alternate close approaches of Mars to the planet earth. In the next chapter this relationship is thoroughly explained.

Chapter 2

The 4½ Year Market Cycle

Up until early October 1987, it seemed as if every investor's fondest dream had finally come true. One of the longest bull markets in history was still in progress, having risen by more than 150% from the dog-days of 1982. Everyone was euphoric. A whole new crop of young analysts and salesmen had come of age without ever having experienced the savaging that a panic sell-off of major proportions could dish out. To be sure, conventional wisdom recognized that a technical correction was overdue—5% maybe, or even 10%. Sophisticated investors expected that a market retreat of manageable dimensions probably would occur, merely to pave the way for the next great leap upward toward . . . who could say? . . . 4000? Even 5000?

They weren't prepared for what happened on Black Monday. The date was October 19, 1987, and the Dow-Jones Industrial Average lost more than 500 points in a few short hours of trading. Having posted all-time highs around 2900 as recently as the previous August, the Dow was now plummeting into a future that looked more infernal than anything that the young analysts could ever have imagined.

Should anyone have suspected that such a spectacular price collapse was in the offing? The answer is yes. Why? Because the 4.5 year cycle with which the stock market had punctually marked almost all of its post-war lows was coming around again, and a new trough was due to be established prior to the next bull phase.

Here is a quotation from page 68 of my earlier book, *Sun*

Spots, Stars and the Stock Market (Pagurian Press Limited, 1979):

> "The 4½ year cycle which begins in 1983 should end in the typical trough formation at a point between six and twelve months prior to the Martian retrograde station occurring on August 27, 1988, that is, toward the end of 1987."

Toward the end of 1987 does encompass October 19. I was not able to narrow it down to the actual date, of course. There are limits to what astrology can do. . .

Now, what is this 4½ year cycle and how does it tie in with Mars?

To answer the question fully, it is necessary to understand the movement of the red planet through the heavens, as seen from the earth.

The ineluctable laws of planetary motion as expounded by Kepler (himself, like Newton, a student of Astrology) require that a planet further from the sun complete a circuit of that body in longer time than a body closer to the sun. Mars is further out than we are, and takes roughly twice as long to make the round trip. Due to the particulars of the two orbital distances, Mars and the earth line up at closest approach roughly every 2.25 years, the period depending on where, in the eccentric orbits, the alignment takes place.

To clarify what has just been said, the reader's attention is directed to Figures 2 through 5. Figure 2 pictures the orbits of earth and Mars when these two bodies are in closest approach. From the earth, it would appear that Mars and the sun were in opposition at such a time, because one would have to look in roughly opposite directions to find these two bodies. One year later, Mars and the earth would be in the positions shown in Figure 3. Here, the earth has made one complete circuit of the sun, but Mars has moved only a little more than half-way around. This is due to the fact that Mars takes just under twice as long as we do to complete its trip. Figure 4 shows the situation another year later. Earth is again in the same position (of course) and Mars has moved just past its original position in Figure 2. Note that these two bodies are not yet in alignment in Figure 4. The earth must still travel

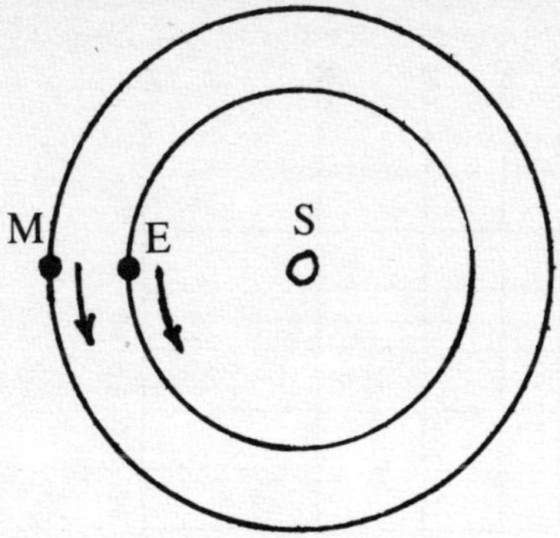

Figure 2

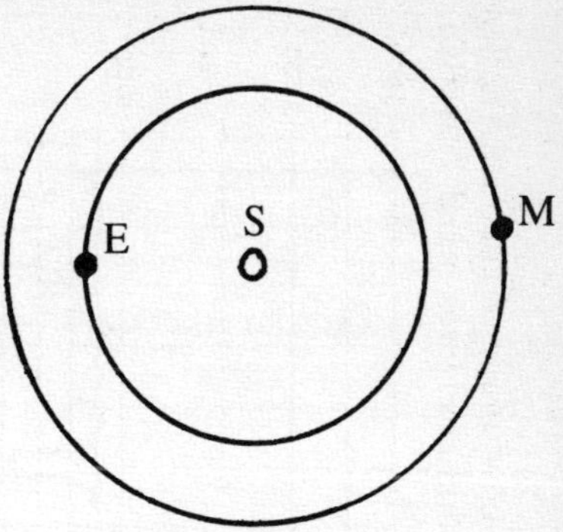

Figure 3

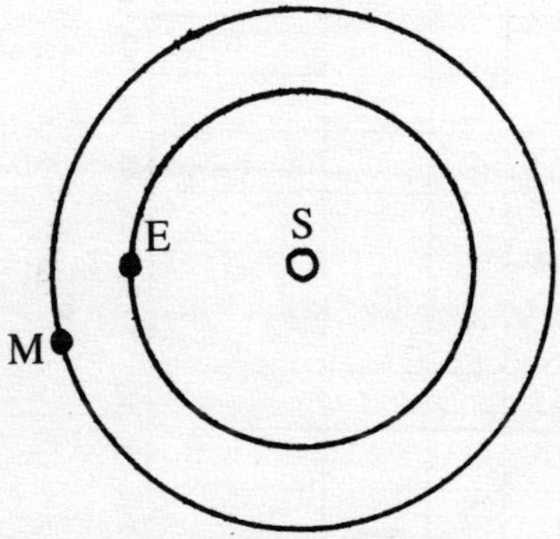

Figure 4

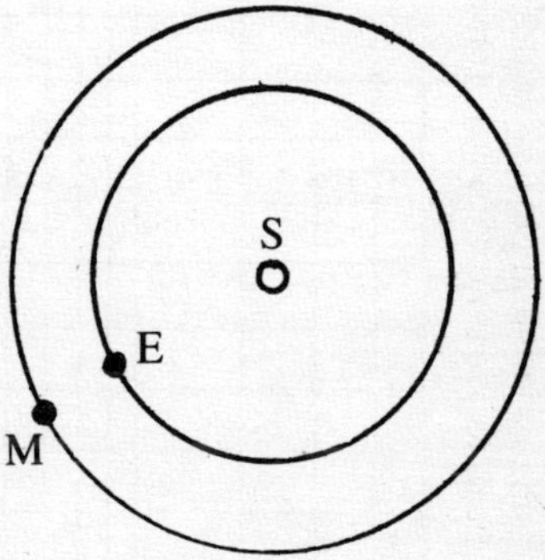

Figure 5

through a small additional arc to "catch up" with its neighbour. Figure 5 shows the actual next alignment position, and it will be noted that this alignment is not at the same location as that in Figure 2. The earth must make a little more than two complete circuits of the sun before it overtakes the red planet again. As stated earlier, the average time between these earth-Mars alignments is about 2.25 years.

Now every time the earth speeds past its alignment position with Mars (Figs. 2 and 5), the phenomenon of "retrograde motion" takes place. Because the earth is moving substantially

Plot of Dow-Jones Industrial Average using month-end closing values. Arrows show timing of alternate Martian retrograde stations, with time interval back to previous "trough" formation. Note that Figures 6 and 7 overlap.

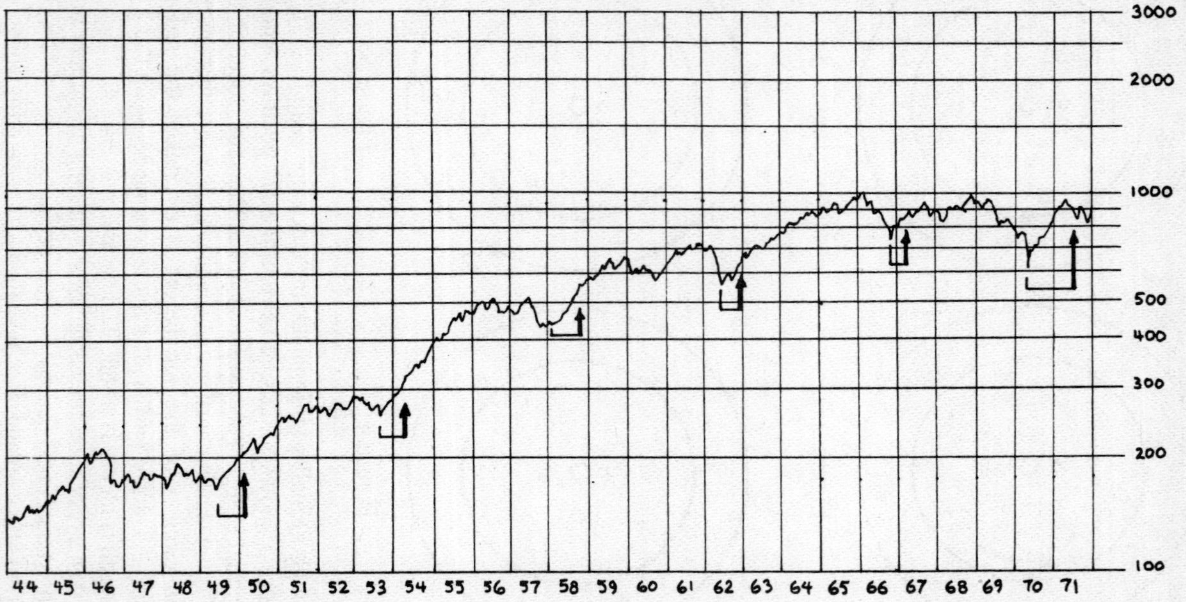

Figure 6

Plot of Dow-Jones Industrial Average using month-end closing values. Arrows show timing of alternate Martian retrograde stations, with time interval back to previous "trough" formation. Note that Figures 6 and 7 overlap.

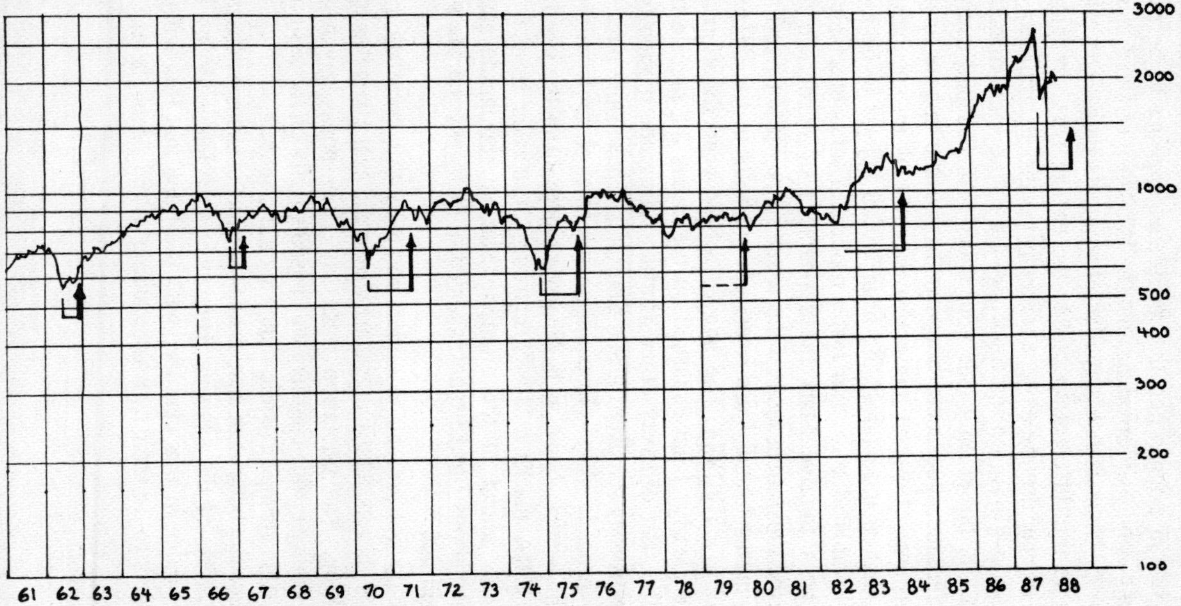

Figure 7

faster than Mars, there is a short period of time bracketing the alignment point when the red planet *appears* to be moving backwards, as viewed against the backdrop of fixed stars. Of course, Mars is not actually retracing its path. The effect is due to the different speeds of the two planets. Imagine you are in a car driving along a roadway parallel to a railway track. There is a train on the track moving in the same direction as you are, but at only half your speed. As you pull even with the engine, you look sideways to see the engine against a panorama of distant mountains. If you took into account only the distant mountains, the train would appear to be moving backwards, due to your faster motion. That is precisely what happens to give Mars the appearance of retrograde (backwards) movement against the much more distant fixed stars.

Figures 6 and 7 are overlapping charts of the DJIA from 1942 to early 1988. The "waves" of 4½ years duration can be seen clearly. I have marked on the charts the precise timing of the start of every other Martian retrograde motion (called in astrology its "Retrograde Station"). These time points usually precede Mars' closest approach to earth by from three to eight weeks. Now, it is clearly evident from the charts that the DJIA marks a "low" or trough formation from 6 to 12 months ahead of the alternate Martian retrograde stations, hence about 7 to 13 months ahead of every second Mars-earth alignment.

I do not pretend to understand why this correlation takes place, but its relative regularity over the past 10 cycles is most striking. The quotation from my 1979 book, given at the beginning of this chapter, was based strictly on the assumption that this strange cosmic dance would continue.

For the benefit of investors, I set out in the next chapter an overview of the Martian patterns and cycles occurring in the remainder of this century, with the hope that this will allow readers to do their own forecasting in the years ahead.

Chapter 3

Stock Market Activity to 2000 A.D.

In light of the correlations elucidated in the previous two chapters of this book, it is appropriate now to inspect the planetary movements over the years which still lie in the future at this writing (July 1988), and to suggest what impact the various Martian patterns and stations may have on the stock market.

In chapter 2 it was pointed out that the "Great Crash" of October 19/87 was right on cue in terms of the 4½ year Martian cycle, according to which it was expected that the DJIA would mark some recognizable trough or low point toward the end of 1987. I will therefore assume that the market levels touched in the period from mid-October to about the end of 1987 represent the end of one cycle and the likely beginning of the next. If that assumption is correct, then we must look ahead two Martian synodic periods (i.e., two close approaches of Mars) to find the next likely trough or low point.

The Martian Retrograde Station timing the low of October 1987 occurs August 27, 1988. It will be recalled that the troughs tend to be established from 6 to 12 months ahead of alternate Retrograde Stations of Mars.

The next Retrograde Station of the red planet occurs on October 20, 1990, being the alternate station (the one not triggering a trough formation). However, the following retrograde station occurring on November 28, 1992 can be expected to produce a market low or trough in the period from about November 1991 to May 1992 (roughly 6 to 12 months ahead of the station).

Let us now look at the 4½ year period between the trough

points, to see whether any triple planet patterns occur that might signal a drop or stall in the Dow-Jones. Specifically, what we are looking for are conjunctions of Mars with one of Jupiter, Saturn, Uranus and Neptune, such that the conjunction is simultaneously in square or opposition to another of these planets, the orb for a square being 5 degrees 20 minutes, the orb for an opposition being 8 degrees.

Around February 22/88, Mars passes the conjunctions to Saturn and Uranus, and then on March 7/88 it conjoins Neptune. However, none of these conjunctions is afflicted by a square or opposition from another of the heavy planets. Indeed, the first two conjunctions are so close together they can be regarded as a triple conjunction, with Mars, Saturn and Uranus all together at the beginning of Capricorn. Interestingly, this triple conjunction is simultaneously in trine relationship with Jupiter at the end of Aries. Although I was not able to find in my study of past performance a statistically significant relation between such a pattern and the DJIA, one can speculate that the positive effect of the trine from Jupiter might be beneficial for the market averages, particularly since we are looking for a new 4½ year wave to be initiated at about this time.

The next conjunction of Mars is that with Jupiter, occurring on March 10/89. This conjunction is not afflicted by a square or opposition from Saturn, Uranus or Neptune.

In February of 1990, Mars again comes around to Capricorn, and passes through the conjunctions of Uranus, Neptune and Saturn (in that order). The conjunction with Uranus occurs on Feb 9/90 at 7 deg 57 Capricorn, when Jupiter at 1 deg 11 of Cancer is in opposition with an orb of less than 8 degrees. This clearly qualifies as a pattern with a value of -2 in accordance with the criteria developed in Chapter 1 of this book. As the pattern comes close to the mid-point of the 4½ year cycle then in progress, it becomes a definite candidate (probability 70%) for the trigger point that turns the bull market into a bear (or that pushes a flat market into a slump). This is all the more significant in view of the fact that Mars is involved with an opposition to Jupiter (see the summary of conclusions at the end of Chapter 1).

On June 14/91, Mars conjoins Jupiter at about 11 deg 17 of Leo, while being in close opposition to Saturn at about 6 Aquarius. The orb is definitely within 8 degrees, giving the pattern a value of −2 in accordance with the criteria developed earlier. It will be recalled that we can probably expect the next trough or low point in the market to occur sometime between about November 1991 and May 1992, assuming that the 4½ year cycle continues to be maintained. If this is the case, then the triple Martian pattern of June 14/91 could well precipitate the final flushing out phase of a bear market prior to establishing the new low. Of course, as happened in the cycle ending in late 1987, the market could conceivably tough it out at high levels to a point just before the timing of the new trough, and then tumble to its low point in a matter of weeks. Nonetheless, the probability is about 70% that the Martian pattern of June 14/91 will turn the market sour.

Between January 29 and February 1, 1992, Mars passes the conjunctions to Uranus and Neptune, the latter planets being still quite close together in mid-Capricorn. However neither of these conjunctions is afflicted by squares or oppositions. Then on March 6/92 Mars conjoins Saturn, still wide of any squares or oppositions.

We may now look at the approximate timing of the subsequent 4½ year cycle, presumed to begin in the period between November 1991 and May 1992. Skipping the intermediate retrograde station of January 2/95, we go to the one occurring on Feb 6/97. Measuring back from that station, we can pinpoint the period from about February 1996 to about August 1996 as being likely to contain the next market low or trough. Thus we have the next wave extending from about early 1992 to about mid 1997, and it is appropriate to inspect this period for possible negative market triggers in the form of afflicted Martian conjunctions with one of the heavy planets.

On September 5/93, Mars conjoins Jupiter at 16 deg 6 of Libra, simultaneously in square formation to Uranus and Neptune, both at about 18 Capricorn. This pattern qualifies for −3 in accordance with the criteria developed earlier, and stands a strong chance of triggering a negative move in the market (70% or possibly better). Note that Mars is square to Neptune, and refer to the summarized conclusions at the end of Chapter 1.

Mars then passes through Capricorn during January of 1994, and between the 16th and 19th of that month conjuncts first Neptune and then Uranus. The latter two heavy planets are less than 1.5 degrees apart at that time. However, neither conjunction is afflicted by a square or opposition from another heavy planet.

A Mars-Saturn conjunction then occurs on March 13, 1994, but without affliction from other heavy planets.

The red planet does not again contact one of the heavies until November 15, 1995, when it conjoins Jupiter at about 19 Sagittarius. Saturn is simultaneously at 18 Pisces, producing a close square that gives the pattern a value of −2 in accordance with the criteria developed earlier. Remember that we are looking for a low or trough to be marked by the Dow-Jones around mid-1997. The Martian pattern just discussed could well be a trigger that begins the downhill slide to the new low.

Mars then conjoins Neptune on December 30, 1995 and Uranus on January 7, 1996, both planets being at the end of Capricorn. However, neither conjunction is afflicted by a square or opposition from another one of the heavy planets.

On March 21, 1996, Mars conjoins Saturn at about 28 Pisces, but this conjunction is unafflicted.

The red planet does not then conjunct another heavy planet until December 15, 1997, which is subsequent to the expected beginning of the next 4½ year cycle beginning mid-1997. The conjoined planet is Neptune at about 28 Capricorn, but the pattern is not afflicted.

The ensuing conjunction with Uranus on December 21, 1997 (at about 7 Aquarius) is also unafflicted.

Mars then conjoins Jupiter on January 20, 1998, again unafflicted by other heavy planets.

On April 1, 1998, Mars comes up to Saturn at about 22 Aries, but no squares or oppositions are present.

The Mars conjunction on November 28, 1999 with Neptune at about 2 Aquarius is wide of a square to Saturn at 12 Taurus, and slightly outside the orb of 5 deg 20 for the square to Jupiter at 26 Aries. Shortly thereafter, on December 13, 1999, the red planet conjoins Uranus at a location (14 Aquarius) which is clearly within orb of the square to Saturn. This complex double conjunction certainly qualifies for at least −2, and I would be very cautious during the 8 to 12 months following the pattern. The probability is at least 70% that the market will enter a sour phase here.

On April 6, 2000, Mars conjoins Jupiter at about 10 Taurus, and is in square to Neptune within our established orb. The value of the pattern is thus −2. Then on April 15 of the same year, the red planet passes Saturn at 17 Taurus, simultaneously squaring Uranus at 20 Aquarius. Again the count is −2. It would be wise to exert caution at this time, especially since the fall-out from the Neptune conjunction of November 1999 may still not have settled.

No further Martian conjunctions with heavy planets occur until after the end of the year 2000.

Chapter 4

The Astrology of Alcoholism

This chapter summarizes an article I prepared in the early 1970's setting out my findings based on a respectably large sample of birth-dates for individuals who had an alcohol-related problem. The article was never published.

My research on the data collected did not reveal that the planet Mars had a necessarily pre-eminent role in the Astrology of alcoholism, although some Martian patterns were clearly in evidence. Nonetheless, the publication of this book seemed an excellent opportunity to present a summary of the article, in view of the research-oriented material offered in the previous chapters.

The data on which the article was based consisted of the date, year and place of birth, as well as the sex, of all of the individuals who, during the calendar year 1967, presented themselves voluntarily to the several treatment centres operated and funded by a government-sponsored organization in a large North-American city, the organization being set up specifically to study and treat cases of alcoholic and drug addiction. There were 337 cases in all. About 16% of the cases were female. Although the organization from which this data was obtained spends a good deal of its treatment and research time on cases of addiction involving drugs alone, all of the cases making up the present sample had a self-recognized *alcoholic* problem, and may or may not have been involved with drugs at the same time.

There are several peculiarities about this particular sample which must be borne in mind. First, while all of the individuals in the sample had an alcohol-related problem, they were *not*

		MOON SIGN											
	AR	TA	GE	CA	LE	VI	LI	SC	SA	CP	AQ	PI	totals
SUN SIGN AR	3	1	1	2	4	4	1	7	1	2	3	8	37
TA	2	0	0	1	3	2	2	1	1	2	1	4	19
GE	0	1	0	1	3	3	5	3	6	2	2	3	29
CA	1	1	0	3	2	2	5	0	3	4	3	9	33
LE	3	5	5	2	3	5	1	2	4	4	2	2	38
VI	1	1	3	3	3	0	3	0	2	2	3	4	25
LI	3	3	5	3	2	1	2	3	1	1	6	0	30
SC	1	2	2	2	4	0	2	0	1	1	2	1	18
SA	6	3	3	3	1	8	1	1	2	5	1	3	37
CP	2	4	0	0	0	4	1	2	1	3	0	4	21
AQ	3	3	2	2	2	4	1	0	1	3	2	1	24
PI	1	1	4	0	4	2	2	3	1	2	2	3	25
totals	26	25	25	22	31	35	26	22	24	31	27	42	

Figure 8

necessarily alcoholics in the medical sense. Second, *they themselves* recognized the problem, which is not always the case for those addicted to drugs or intoxicants. Third, these individuals all came *voluntarily* to a treatment centre for help, which again is not something that all those with alcoholic problems would necessarily do.

There are thus three known factors that characterize each case in this sample. This presents the problem of deciding to which of these we should tie a given Astrological pattern which appears to crop up significantly more or significantly less times than chance would allow. Here is an example: The simplest astrological breakdown of the data is by sun-sign. Figure 8 is a table which includes this information. The numbers along the right edge give the frequency of occurrence for the various sun-signs, while the numbers along the bottom give the moon-sign frequency. The table also shows all 144 sun/moon combinations. (Students of this subject will realize that an individual birth-chart with the sun located in a given one of the zodiac signs can at the same time have the moon located in any of the 12 signs. It simply depends on the precise day and hour of birth.) Now with a total of 337 cases, one would expect about 28 (one-twelfth) to fall into each sun-sign division. However, there are only 18 Scorpio suns. If we agree that the deviation from the expected occurrence is a significant one, then we are in some difficulty to decide whether the low score for Scorpio is due to a lower addiction rate because of the traditional Scorpionic self-control, or is due to a strong disinclination on the part of Scorpios to reveal problems to others because of the traditional secretiveness of this sign. Unfortunately, there is little we can do to escape from this kind of dilemma in the present study. Nonetheless, the findings remain most intriguing.

Let us begin by making a closer examination of Figure 8, which as mentioned shows the frequency of occurrence for the sun-signs and the moon-signs. It also shows the occurrence of the various sun-moon combinations. Because the birth data did not include the precise hour and minute of birth, the breakdown by moon sign is not 100% reliable. What I did was to determine the sign holding the moon at noon local time for the date and place of birth. Because the moon moves roughly 12-15 degrees per day, the arbitrary selection of the moon position as that at noon local time could introduce an error of up to 7½ degrees. This may mean that some moon-sign designations are incorrect, however I believe the error will be relatively inconsequential.

Even allowing for a margin of moon-sign error in the table of Figure 8, certain striking departures from chance expectation are worth pointing out. Firstly, if we were to distribute our sample (337) uniformly among the 144 slots for the different possible sun-moon combinations, we would expect about 2⅓ instances for each one. Note that the combination with the highest incidence is Cancer sun with Pisces moon, there being nine cases with this arrangement—almost four times the "chance" frequency! The next two most heavily tenanted combinations are Aries sun with Pisces moon and Sagittarius sun with Virgo moon, both holding eight cases (more than three times the expected number). Note should also be taken of Aries-Scorpio, Gemini-Sagittarius and Sagittarius-Aries.

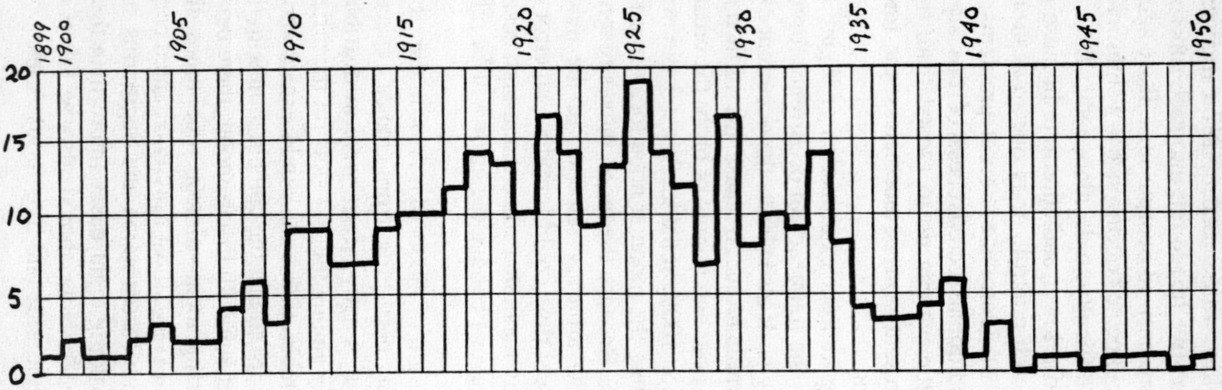

Figure 9

1	2	3	4	5	6	7	8	9	10	11	12	13	14	15	16	17	18	19	20	21	22	23	24	25	26	27	28	29	30	31	DATE OF MONTH
13	17	13	4	10	6	6	16	11	15	9	14	5	8	15	10	15	11	10	13	14	12	10	8	15	13	9	11	8	10	8	TOTAL

Figure 10

It is of interest that, generally speaking, moon in Pisces seems to crop up with excessive frequency, and indeed occurs almost as many times as the other two water signs (Cancer and Scorpio) put together. As to sun signs, it will be observed that Aries, Leo and Sagittarius tend to dominate the others in frequency (these being the fire signs), and that Scorpio, Taurus and Capricorn (in that order) appear with the least frequency.

Another intriguing finding is that the so-called square and opposition aspects between the signs containing the sun and moon in a chart do *not* appear to dominate the other aspects in frequency, as certain teachings in traditional astrology might suggest. It is considered "difficult" or "problematic" to be born with a square or opposition between the Lights, which means that (for example) the combination of Aries sun with Capricorn or Cancer moon (the square) or with a Libran moon (the opposition) might be expected to produce the kinds of life-tensions that could lead to an alcohol problem. In actual fact, the lunar positions of Pisces and Scorpio constitute almost 41% of the total sun-in-Aries combinations, these signs being 30 degrees and 150 degrees, respectively, away from Aries.

Some sun-moon squares and oppositions do occur with higher than expected frequency, and these are worth mentioning. A list follows:

Gemini sun with Sagittarian moon (6)
Cancer sun with Libran moon (5)
Scorpio sun with Leo moon (4)
Sagittarian sun with Virgo moon (8)

Before plunging further into this analysis, it is worthwhile to note the distribution by birth-year of the cases making up the sample. This distribution is shown in the histogram of Figure 9, from which it will be noted that the year holding the greatest number of cases is 1925. As previously noted, the year in which all of these individuals presented themselves voluntarily for treatment was 1967, a time-span in which those born in 1925 would have had their 42nd birthdays. It is well known in astrology that the time around the 42nd birthday

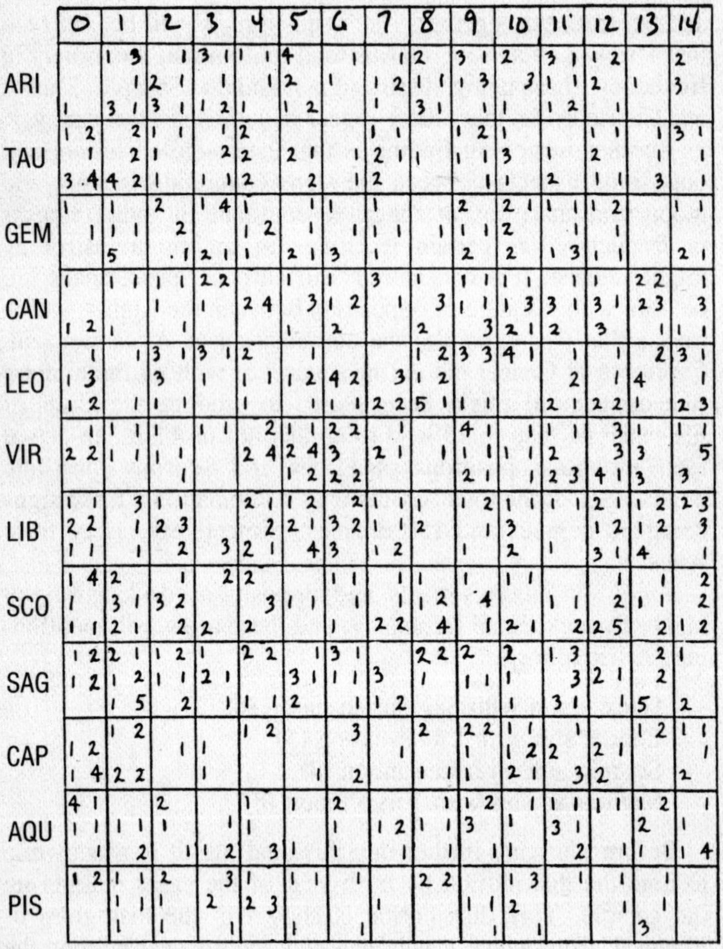
0 1 2 3 4 5 6 7 8 9 10 11 12 13 14
ARI
TAU
GEM
CAN
LEO
VIR
LIB
SCO
SAG
CAP
AQU
PIS

Figure 11A

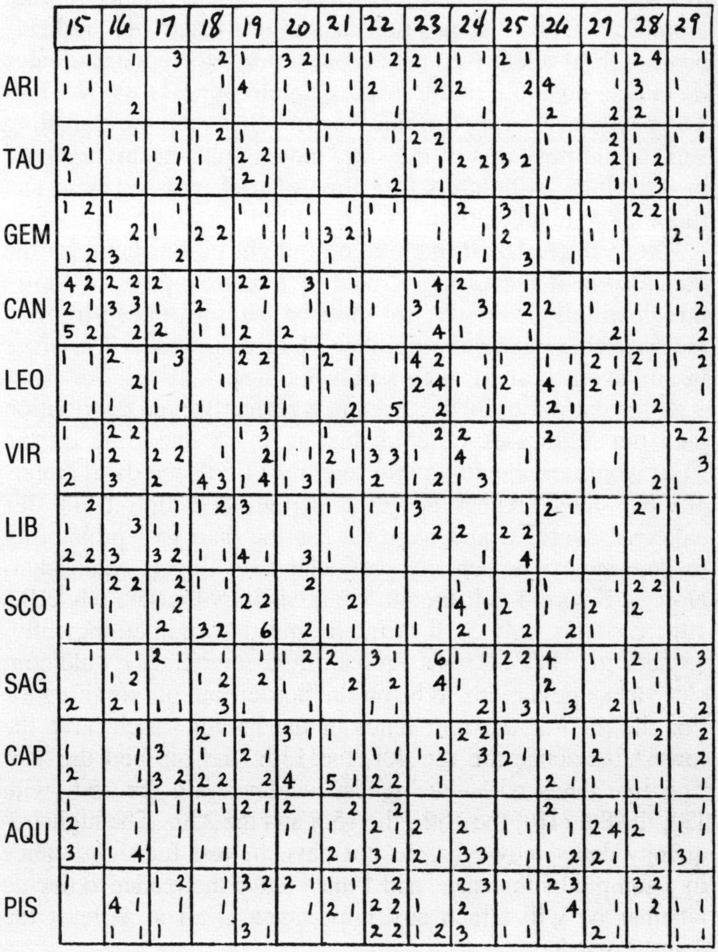
15
16
17
18
19
20
21
22
23
24
25
26
27
28
29
ARI
TAU
GEM
CAN
LEO
VIR
LIB
SCO
SAG
CAP
AQU
PIS

Figure 11B

tends to be a stressful time of life for every individual. As the Astrologer Hirsig has ably pointed out, this is the period when transiting Uranus (the actual position of that planet) comes around to oppose its position in the birth-chart (its "radical" position). At about the same time, transiting Neptune squares its radical position, and transiting Saturn opposes its own radical position. These "difficult" aspects between the transiting and radical positions of the slow-moving planets are regarded in Astrology as pointers to a time of trial, reassessment and (hopefully) regeneration.

Those interested in numerology might be intrigued by the breakdown of the cases by date of month. Figure 10 shows this distribution. Firstly, we must be careful to keep in mind the special status of the number 31 in this breakdown, since in any given year 31 occurs only 7 times.

If the dates 1 to 30 occurred in a perfectly even distribution then our 329 cases (after deducting the 8 occurring on the 31st) would divide down into roughly 11 on each date. Naturally we must expect to get a certain deviation from this expected average, and we must use the theory of probability to determine whether any particular deviation is significant. Note in Figure 3 that the 4th contains only 4 cases, while the 13th contains only 5. It is interesting that, numerologically, 13 adds to a 4 ($1+3=4$). Probability theoreticians would consider these deviations to be on the border-line of 'significant'. For the record, the four dates of the month which have the lowest incidence are the 4th, the 13th, the 6th and the 7th. The five dates with the highest incidence are the 2nd (with 17), the 8th (16), the 10th, the 15th and the 25th. The high-frequency dates, however, do not pass the test for significance in a sample of this size, and I must leave the reader to decide whether he will attach any importance at all to at least the dates 2 and 8.

We next turn to a breakdown of the sample in terms of the zodiac degrees tenanted. Figure 11 is a table which shows the frequency, in each zodiac degree, of the occurrence of the sun, Mercury, Venus, Mars, Jupiter and Saturn for all of the cases. Each degree rectangle has 6 positions where a number may occur. The number of times the sun occurs in a given degree is always shown at top left. Mercury is shown at top right, Venus at middle left, Mars at middle right, Jupiter at bottom left and Saturn at bottom right. Each rectangle indicates the frequency of the degree of arc *following* the number at the top of the column. Thus, a rectangle in the column headed 7 shows planets occurring between 7 deg 00 and 7 deg 59, which means that a rectangle in the column headed 7 actually refers to the *eighth* degree of the particular sign. Taking an example from the table, between 6 deg 00 and 6 deg 59 of Virgo the sun occurs twice, Mercury twice, Venus three times, Mars no times, Jupiter twice and Saturn twice. The positions for the planets in Figure 4 were all taken at Greenwich noon of the date in question, and were not corrected for local noon.

The reader may question why the moon and the three trans-saturnian planets have been omitted from Figure 11. The reason for the omission of the moon is, of course, that we have no way of determining its exact degree position, as we do not know the precise hour and minute of birth for the cases in our sample. The trans-saturnian planets (Uranus, Neptune and Pluto) are omitted because none of them would make a complete circuit of the zodiac in the period containing all of the birth dates in our sample. Indeed, we may not even be justified in including Saturn in Figure 4, because the ringed planet's sidereal period is about 29 years, and our histogram of Figure 9 does not remain relatively constant over that long a period.

The table of Figure 12 is derived from Figure 11, and shows the frequency of occurrence of the planets Mercury, Venus, Mars, Jupiter and Saturn in the twelve signs. A few points concerning Figure 12 are worth noting. Firstly, the planets generally appear with greater than average frequency in signs in which they are considered to be "debilitated", i.e. badly placed. In traditional Astrology, each planet is regarded as having one or more "home" signs which it is said to "rule". Further, each planet has one additional sign in which it is said to be "exalted". For example, Mars rules Aries and Scorpio,

	Merc	Ven	Mars	Jup	Sat
ARI	33	36	23	22	21
TAU	23	29	26	22	28
GEM	22	20	25	29	19
CAN	28	33	41	29	23
LEO	33	32	30	29	33
VIR	25	35	40	37	29
LIB	32	23	31	33	32
SCO	25	26	31	27	40
SAG	38	24	30	27	34
CAP	25	22	21	34	28
AQU	20	34	18	28	38
PIS	34	24	22	21	13

Figure 12

and is exalted in Capricorn. Again traditionally, the signs *opposite* the signs in which a planet is "home" or "exalted" are called the "detriment" and "fall", respectively, of that body. Continuing with the example of Mars, the red planet is in its *detriment* in Libra (opposite Aries) and Taurus (opposite Scorpio), and is in its *fall* in Cancer (opposite Capricorn). Libra, Taurus and Cancer are thus viewed as difficult placements for Mars.

Looking now at Figure 12, it is of considerable interest that Mercury occurs the greatest number of times in Sagittarius and in Pisces, these signs being opposite the ones (Gemini and Virgo) which Mercury is considered to rule. Similarly, Venus crops up with the greatest frequency in the signs Aries and Virgo, which are the detriment and fall, respectively, of that planet. Mars is found the greatest number of times in Cancer, its fall as explained above. Interestingly, Mars does not show up with excessive frequency in Libra, which is traditionally deemed to be its detriment. However, the red planet does occur an excess number of times in Virgo. This would tend to add confirmation to a common view among astrologers (shared by this writer) that Mars is badly placed in Virgo.

Jupiter also occurs excessively in Virgo, and here we are on firmer ground since Virgo is opposite Pisces, over which Jupiter is said to have at least half-rulership. (Back before the telescope era, when scientists and astrologers thought that the *visible* planets were the only bodies in the solar system, Pisces was assigned to the rulership of Jupiter. When Neptune was later discovered, Jupiter was at least partially dethroned in favor of the God of the Sea. However the Jovian affinity to the sign of the fishes still remains.)

By contrast to the other planets, Saturn is an enigma. It occurs with the greatest frequency in the signs Scorpio, Aquarius and Sagittarius, none of which debilitates the ringed planet. The explanation for this distribution, however, may relate to the histogram of Figure 9. For example, throughout 1925, the birth-year containing the greatest number of cases in our sample, Saturn was in Scorpio. Similarly, for all but three months of 1929, a birth-year containing 17 cases, Saturn was in Sagittarius. The planet's excessive frequency of occurrence in Aquarius, a sign which it was anciently considered to rule, is rather a mystery to this writer. Saturn was in Aquarius from March 1932 to February 1935, except for a 3-month period within this span.

It is interesting to look at the planets Mercury, Venus, Mars and Jupiter from a probability point of view. If we take the ancient rulerships of these planets (where Jupiter rules both Pisces and Sagittarius), we find that each of the bodies has its maximum occurrence either in its detriment or in its fall; that is to say, in one of three possible signs out of twelve. In a true random distribution, the likelihood of a planet maximum occurring in one of three signs out of twelve is 1:4 or 25%. But in our sample, *four* planets have maxima in areas of the zodiac in which they are only 25% likely to occur, in a random distribution. Therefore the probability of this particular result is $1/4 \times 1/4 \times 1/4 \times 1/4 = 1/256$. The odds against our result being due to pure chance are 256 to one. In the field of probability as applied to statistical surveys, odds of 20 to one are considered 'significant', while odds of 100 to one (or better) are regarded as 'very significant'.

Attention is once again directed to the table of Figure 11 for a detailed examination. Firstly, an even distribution of the six planets throughout all of the degrees of the zodiac would give us about 5.6 occurrences per degree (taking all the planets together). For each specific planet, an occurrence of about 0.94 per degree is expected. The more interesting accumulations are as follows:

a) Mercury appears six times between 23°0 and 23°59 Sagittarius.

b) Mars occurs eight times between 4°0 and 5°59 Virgo, and occurs five times between 14°0 and 14°59 Virgo.

c) Jupiter occurs seven times between 0°0′ and 1°59′ Taurus (thus, 1/3rd of the total Jupiter-in-Taurus positions occur in the first two degrees of Taurus), five times from 1°0′ to 1°59′ Gemini, five times from 15°0′ to 15°59′ Cancer, five times from 8°0′ to 8°59′ Leo, 20 times from 12°0′ to 19°59′ Libra (thus, 20 occurrences of Jupiter in Libra are located in an 8-degree span), five times from 13°0′ to 13°59′ Sagittarius, and eleven times from 20°0′ to 22°59′ Capricorn (thus, 1/3rd of the Jupiter-in-Capricorn positions take place in a span of 3°). Jupiter occurs five times from 21°0′ to 21°59′ Capricorn.

d) Saturn occurs five times from 22°0 to 22°59 of Leo, ten times from 18°0 to 20°59 of Virgo (thus 1/3rd of the Saturn-in-Virgo positions occur in a span of 3°), five times from 12°0 to 12°59′ Scorpio, and six times from 19°0 to 1°59′ Sagittarius.

Generally speaking, there is an excessive accumulation of planets (taken all together) in the areas of 15°-16° Cancer and 5°-6° Virgo.

In regard to these degree accumulations, it must be stressed that some of them may be accounted for at least partially by the fact of planetary stations in or near the degrees mentioned. On the other hand, over the period during which the births took place, a great number of planetary stations occurred which do *not* show up with noticeable excess frequency.

Aspects were also studied, and the orb used was 9°. The principle aspect noted was Neptune conjunct Mars, which cropped up 27 times using the 9° orb, as against an expected frequency of about 17. This result represents about 2.5 Standard Deviations, for which the odds against a chance occurrence are better than 100 to 1. It is not surprising that this particular aspect should crop up an excess number of times, in view of the traditional linking of Neptune with 'escape' of all kinds. Mars provides the energy to deal with and function in the real world, and its conjunction with Neptune could point to an undermining or debilitation of that energy source, leading to frustration and the desire to escape from the conflicts it engenders. Interestingly, the opposition and square aspects between these two planets show up less times than chance would suggest, while the trine and sextile occur slightly more often, as the following table shows:

Aspect Mars-Neptune	**Chance Expectation**	**Actual Occurrence**
Conjunction	17	27
Opposition	17	15
Square	34	31
Trine	34	36
Sextile	34	28

Summarizing the astrological indicators thus far obtained, we may say that the following factors tend to crop up with greater than expected frequency in a relatively large sample of cases having alcohol-related problems:

Sun-Moon Combinations
Sun in Aries, moon in Scorpio
Sun in Aries, moon in Pisces
Sun in Cancer, moon in Pisces
Sun in Sagittarius, moon in Virgo

Sun Sign
The fire signs have a greater than expected frequency (while Scorpio, Taurus and Capricorn have less).

Moon Sign
The moon in Pisces is clearly the lunar position with the greatest frequency.

Signs of the Planets
Mercury in Sagittarius and Pisces
Venus in Aries and Virgo
Mars in Cancer and Virgo
Jupiter in Virgo

Degree Locations
Any planets around 16 Cancer
Mercury at 23-24 Sagittarius
Mars at 4-6 Virgo and 14-15 Virgo
Jupiter at 0-2 Taurus, 15-16 Cancer, 8-9 Leo, 12-20 Libra, 13-14 Sag., and 20-23 Capricorn.
Saturn at 12-13 Scorpio, 19-20 Scorpio, 22-23 Leo, 18-21 Virgo and 1-2 Sag.

Date of Month
The most heavily tenanted dates (in decreasing order of frequency) were: 2, 8, 10, 15 and 25.
The least heavily tenanted dates were 4, 13, 6 and 7.

Aspects

The most often occurring aspect between planets was the Neptune-Mars conjunction.

Conclusions

While this study cannot be used to derive a foolproof method of detecting alcoholic proclivities in a birth-chart, it nonetheless furnishes a number of confirmatory indicators of the addictive personality that may be employed where the remainder of the chart establishes primary evidence of that weakness. The study also tends to confirm much of what traditional Astrology has long contended, particularly in regard to sign rulerships and debilities of the planets.

Chapter 5

Mars and Armed Conflict

An interesting and statistically significant correlation can be identified, linking the Direct Station of the planet Mars with outbreaks of military conflict on the earth. For centuries Astrologers have asserted that the red planet was related to war in some way, but they were not able to demonstrate the correlation to the satisfaction of hard-headed skeptics. Despite the obvious mythological and linguistic associations with the name of the planet (martial arts, martial law, etc), the war implications of Mars have thus far gone unrecognized.

But now, with all of the military activity that has taken place since the outbreak of the Second World War, it is possible to examine this fascinating correlation using enough instances to establish a statistically significant relationship. The "Mars Connection" can be clearly seen in the chart of Figure 13. Mars makes its closest approach to the earth roughly every 26 months, a period that varies slightly due to the eccentric orbits of the two planets. The chart shows a base line length of 26 months, beginning at the Martian Direct Station and ending at the next subsequent Direct Station. Figure 13 is drawn as if all the station-to-station periods were superimposed. Below the base line are marked numbers identifying the beginnings of the major outbreaks, as tabulated below:

Direct Station	Outbreak	Event	Time Lapse
1. Aug 24/39	Sept. 1/39	World War II begins	7 days after
2. Nov. 11/41	Dec. 7/41	Pearl Harbour attacked	4 weeks after
Jan. 10/44		(War continuing)	
Feb. 21/46		no obvious event	

3.	Mar. 29/48	Apr. 1/48	Berlin Blockade	3 days after
4.	Mar. 29/48	May 14/48	First Arab-Israeli war	6 weeks after
5.	May 3/50	Jun. 25/50	Korean War begins	8 weeks after
	Jun. 10/52		no obvious event	
	Jul. 29/54		no obvious event	
6.	Oct. 11/56	Oct. 23/56	Hungarian Revolution begins	12 days after
7.	Oct. 11/56	Oct. 29/56	Suez invasion	18 days after
8.	Dec. 20/58	Jan. 1/59	Cuban revolution, Castro takes power	11 days after
9.	Feb. 6/61	Apr. 17/61	Bay of Pigs invasion	6 weeks after
	Mar. 16/63		no obvious event	
10.	Apr. 19/65	Apr. 24/65	Civil war in Dominican Republic	5 days after
11.	May 27/67	Jun. 5/67	6-day war (Israel)	8 days after
	Jul. 8/69		no obvious event	
	Sep. 9/71		no obvious event	
12.	Nov. 26/73	Oct. 6/73	Yom Kippur War (Eg. Syr. Isr.)	7 weeks before
	Jan. 20/76		no obvious event	
13.	Mar. 3/78	Early Mar.	Israel occupies S. Lebanon	a few days after
14.	Apr. 6/80	April 80	Israel again occupies S. Lebanon	a few days after
15.	May 11/82	May 21/82	Br. forces land on Falklands—fighting begins	10 days after
16.	May 11/82	Jun. 6/82	Israel invades Lebanon as far as Beirut	4 weeks after
	Jun. 19/84		no obvious event	
	Aug. 12/86		no obvious event	

The above table includes all 23 Martian Direct Stations from the beginning of WW2 to the present writing (July, 1988), 13 of which (57%) are followed closely by one or more outbreaks of armed conflict (16 outbreaks in all).

The most surprising result however emerges when one looks at this correlation statistically. If we arbitrarily take a time "window" 4⅓ months long, beginning 7 weeks ahead of the Direct Station, we will have a span equal to one-sixth of the total base line, and within this span all of our events fall. To have all 16 events occur in the same sixth of a recurring cycle has the same probability as throwing a die sixteen times

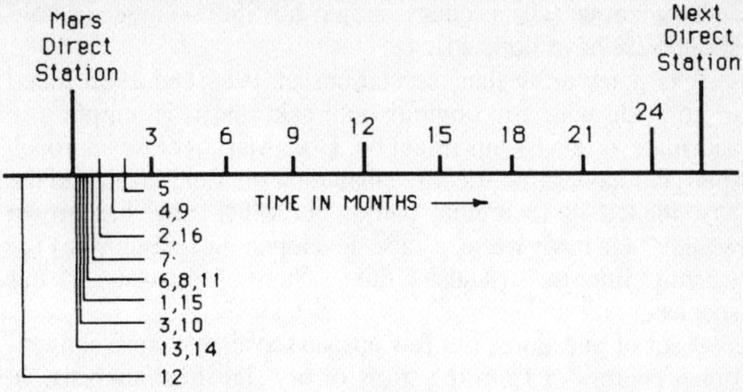

Figure 13

and having it come up the same number every time. For the mathematicians, the chances of this happening are $\left(\frac{1}{6}\right)^{16}$, i.e. $\frac{1}{6}$ multiplied by itself 16 times. The likelihood is only one in 2.8 trillion!

It must be pointed out that the occasional war or coup does take place outside the 1/6th window we have established. WWI is one example. The Iran-Iraq conflict beginning Sep. 22/80 is another. However the overwhelming proportion of such events falls squarely within the window timed by the Martian approach and Direct Station.

Although the statistics clearly establish that a correlation of some kind is at work, they do not throw any light on the *mechanism* by which Mars might exert an influence on mankind's actions. Until scientists can pinpoint an effect of this kind, the "Mars Connection" must remain simply a statistical curiosity.

I will shortly set out the precise dates of all Martian Direct Stations between 1988 and the end of the century, to allow readers to keep track of developments for themselves. Before doing so, however, I wish to stress again that although almost all armed outbreaks since the beginning of the Second World War have tended to take place within the "Mars windows" defined by close approaches of the red planet to the earth, the converse is not true. In other words, we cannot say that every such window will produce major hostilities (indeed, only about 57% have done so).

It is noteworthy that the stations of 1984 and 1986 failed to coincide with any obvious outbreak, and it is tempting to speculate as to why this might be. Does Mars at close approach (perigee) have some specific emanation or "field" which in the past incited us to mutual battle, but which now has grown weaker? Or have we as a race developed an immunity to the warring impulse? (And is this a sign of maturity in our species?)

A lot of questions, but few answers available from conventional sources or from the study of dry statistical analysis. In the hope that the reader will find the remarks of an *unconventional* source of some passing interest, the next chapter contains a commentary on Mars by my "guide" Hilarion, who thus far has dictated some 16 books on esoteric and related topics (all published by Marcus Books). Hard-headed skeptics may do with this information what they wish. For a general discussion regarding channeled material, see Chapter 9.

Here are the dates of all Martian Direct Stations to the year 2000:

October 28/88
January 2/91
February 15/93
March 24/95
April 27/97
June 4/99

Chapter 6

Hilarion on Mars

The planet you call Mars has been very instrumental in quickening the level of spiritual insight among mankind over the past centuries. As we have explained in our book, *The Nature of Reality,* the red planet became the host of a gigantic and powerful "thought-form" of anger and hostility which was left over, so-to-speak, after the disintegration of a planet which had previously existed between the orbits of Mars and Jupiter. The destruction of that body produced the asteroid belt, which has been something of a conundrum to scientists who did not understand the origin of the phenomenon.

The thought-form of anger and hostility had been created on that previous planet by the actions and thought-patterns of beings who incarnated on its surface for a period of several millions of your years. By continually warring with each other and often attacking races on other planets, they managed to build up a powerful aetheric energy around their planet, which reflected the same dark and destructive urges which characterized the beings themselves. Now, since aetheric energy is not destroyed along with the physical matter with which it is associated, the disintegration of the planet created a serious question as to what should be done with the energy. The very lofty Beings who guide the destinies of planets were faced with the task of "binding" the aetheric form in some way that would not allow it free access to the less mature races on nearby bodies, including humanity on this planet. If the form had been allowed to focus itself on the earth at the time in question, the fledgeling race of mankind would in all probability have allowed it to transmute into anger and hatred the natural and healthy assertive instincts which the race shares with most animal species.

For this reason, a decision was taken to tie the aetheric energy to the planet Mars, allowing it to have unfettered access to humanity only at the times of Martian close approach, occurring at an interval just over 2 years in length. More specifically, a decree of sorts was handed down, according to which the transfer of negative influence from Mars to the earth could occur only in the time "window" defined by the apparent retrograde motion of the red planet. This is why the outbreaks of armed conflict described by our channel in the previous chapter tend to group themselves in the weeks or months immediately following the time of retrogradation. During retrograde motion, the energy is assimilated by those humans who are open to the darkness it represents, but they usually do not act upon the impulses until after Mars has resumed its direct or forward motion. This is due to the fact that, astrologically, whenever a planet is in retrograde motion, its essence is not easily expressed at the physical or practical level. Thus, retrogradation permits the "stock-piling" of Martian energy by those attuned to violence and anger, but it represses the actual manifestation of that energy.

The fact that humanity has withstood the urge to large-scale violence which Mars exerted in the middle 80's does suggest that improvement of a kind has been attained. The power of the thought-form associated with the red planet has not, however, diminished in any sense. Indeed, mankind has unwittingly built up around *its own* planet a similar (though less powerful) thought-form of anger, malice and contempt. This has occurred primarily during those very outbreaks of violence which were precipitated by the proximity of Mars at various times in its orbit.

We cannot assure the reader that future visitations of the red planet will be equally well withstood. Part of the reason for the absence of clearly discernable outbreaks in '84 and '86 was the fact that the superpowers were preoccupied with keeping vassal states in line, and indeed were somewhat aghast at the fury with which the lesser nations (particularly in the middle east) were ready to savage each other. The major players on the world scene give the appearance of a softening of attitudes, and with further work on the part of those who truly wish for a peaceful planet, a complete reconciliation can be achieved. However, there are factions in both blocs that seek only their own self-interest, and if these should carry the day, then you may expect the earth to witness at least one episode of mass destruction before the century is out. If that should transpire, you will see the phenomenal transforming effect it will have on the collective psyche. Indeed, though it sounds a heartless thing to say, the dropping of a single nuclear bomb on a populated center would do more to spread the milk of human kindness than a thousand petitions, a thousand marches and rallies, could ever accomplish. We pray that such a demonstration of the folly of war will never be necessary. But the decision is in the hands of mankind—hands that have always held the key to the future of this planet.

Chapter 7

Hilarion on Money and Avarice

The publication of this book affords us a special opportunity to address individuals who would not normally be led to look at our other writings, and we do not wish to let the chance go unused.

Many of the readers of this work were drawn to it by the lure of personal profit, the thought being that if there were indeed some kind of cosmic two-step between Astrology and the stock market, the canny investor should not allow the guffaws of his fellows to disuade him from learning the tune. There is nothing whatever wrong with that approach. It is not a sin to want to be successful, or rich, or famous–or to achieve any other station in life. You were placed on this earth to explore, play and learn, in the hope that you would finally discover *yourself*, your own true essential being. If, along the way, you should happen to pick up the ownership to Boardwalk and Park Place, requiring that the other players regularly pay you more rent than you do them, well that is merely a fact of economic life. The important thing is not how much money you manage to accumulate before the body gives out, but rather your *attitude* toward the money you have. There is an old adage among the sages of the earth, according to which "nothing a man does (or has) is as important as what he *thinks* about it".

Consider the ancient practice of renunciation: it used to be thought that, by giving up all of one's wordly goods, one could achieve nirvana, enlightenment, godliness. Yet many of the ascetics who lived lives of complete renunciation are now reincarnated again as stockbrokers and investors looking for a way to make a "killing on the market". Did holiness elude them in that previous experience? Not necessarily. But they may have died to that life without having entirely extinguished every spark of acquisitiveness from the soul, and are now back again to allow that ember to burst into full flame. How else can they realize that it is there still? You see, renunciation is often merely a way to "put aside" a particular lesson or experience until it can be dealt with more expeditiously. In the desperate poverty of 19th century India, for example, an individual would find little opportunity to learn the lessons that come with riches. But today! With a stock market in which a fortune can be made almost overnight (and as quickly lost)—what better setting to allow all the acquisitive desires to surface?

Do not read into the foregoing paragraph any condemnation of wealth. It was mankind who dreamed up the notion that to accumulate excessive riches was wrong—or at least an impediment to achieving higher ground in a spiritual sense. But the Power that created your world passed no such law. Indeed, to impute a disapproval of wealth to a Creator Who has placed such an abundance of it into the garden for His children seems intuitively contradictory, does it not? What we are saying is that the *attitude* with which wealth (or its absence) is contemplated is the key factor—not how many digits your bank balance may have.

There are many attitudes toward money that do, however, act as impediments to achieving that fine balance of internal traits which humanity must achieve within the next decades, if the marvels of the coming century are fully to flower. Let us deal briefly with a few of these.

The primary flaw connected with money is that of allowing material wealth and its acquisition to displace other human values. Everyone is familiar with the Scrooge complex, thanks to the wonderful Dickens story, and it is instructive to examine that tale in the light of a more spiritual perspective. Of course, every reader will, without hesitation, condemn Ebeneezer's initial attitude, which led him to begrudge Cratchett even Christmas day as a holiday. But there lies within many people an echo of that same viewpoint, impelling them to sacrifice human warmth and love on the glittering altar of material possessions. The premier instance of the Scrooge complex in the present day is the father who neglects to spend time with his children because he is too busy accumulating money. Typically, the transgression is far less egregious than the wrongs of which Scrooge was guilty, but of course Dickens was a master at using exaggeration to make a point.

Another defect of the human attitude toward wealth relates to quantity. The current civilization is, unfortunately, one in which wasteful consumption is encouraged. Manufacturers make products with "built in" obsolescence, and it is a rare item that outlasts its owner's interest without requiring repair. This circumstance has fostered a conviction that one needs more and more cash merely in order to stay in the race, as it were. It is assumed that the car, the television and similar items will all have to be replaced at short term, or alternatively be repaired and maintained at substantial cost. Yet it would have been possible, by using superior materials and manufacturing techniques, to make these products much longer-lasting. It is sobering to realize that the attitude of manufacturers, so well depicted in the British film, *The White Suit,* will not change until a replacement for the profit motive is found.*

The approach to money has been further distorted due to the *glorification of self* which has emerged particularly in the twentieth century. While humanity's guides intended and expected that a strong trend toward individualization would occur, it was hoped that the associated inclination toward vanity and self-absorption could be kept in check. It is part of mankind's path to explore the many facets of each individual's makeup, to hone special talents, to develop the delightful quirks and idiosyncracies that make each person unique. But vainglory and egotism are stumbling blocks to spiritual advancement. They can never lead a soul to the higher ground of true enlightenment, nor provide a key to unlock any door. And most unhappily, these traits are whipped into a frenzy by

---

* One is suggested in Hilarion's book, *Vision.*—Ed.

the availability of countless items for personal adornment, prestige automobiles for conspicuous consumption, and lavish houses to announce the self to the world. Yet there is a light within all this darkness: it is the opportunity to *test* oneself against the temptation to accumulate "things" like an encrustation about the ego. If you can walk serenely amid the siren calls of avarice and conceit, turning gently away from vainglory while continuing to develop and foster the *internal* riches with which the Creative Power has endowed you, then for once and for all you can be free to soar on wings of your own making.

Then there is the "I'm poor but I'm spiritual" attitude toward the question of money—an attitude which is held by many seekers and New Age thinkers. Let us say this as emphatically as we can: *it is not spiritual to be poor!* It is also not spiritual to be rich or black or white or Hindu . . . or any other category of human existence. None of these qualities guarantees that the celestial mountain can be scaled. The only way to be spiritual is to *be spiritual!* Indeed, the person who can keep his focus on higher truths while at the same time plunging fully into the world of practical achievement—making a contribution to society and taking gracefully the rewards which society offers for that effort—is the one who has truly overcome the temptations of the material plane. It is easy to triumph over avarice in the hermit's cave, or by taking the monastic vows, or by drowning in a sea of cheap wine on skid row . . . It does not come as readily in the midst of plenty.

What must be understood here is that the race of mankind has come at last to face the final lesson of materiality: it must learn to administer the richness of creation for the benefit of *all* life-forms; it must become the wise and gentle caretaker of the earth garden. The "poverty mentality" has no place in this new scheme of things. Humanity will become fully conscious of the wealth of reality and the endless abundance which the Creative Power has provided, and will delight in its role as custodian of the planet.

It is lamentable that current modes of thought foster what might be called a "scarcity syndrome" among many people.

The false view has arisen that there is only so much of any given commodity, and that the more one person has, the less remains for others. This attitude is particularly in evidence in regard to money itself. In actual fact, the banking system in all western countries is capable of expanding the money supply to accommodate any increase in national output. All that is required is greater effort and ingenuity on the part of individuals. Even more unfortunate is the tendency to apply the scarcity attitude to areas where it is totally inappropriate. Consider the young child who is jealous of the attention and love which a parent is showing to his younger brother or sister. The child has the mistaken view that if little brother or sister gets a certain quantity of love, there will be less for him. How sad that love, the very source and engine of all that you have in your world, should come to be regarded as a glass that could be emptied. Do not fall into the error of imagining that your material world is one of lacks and limits! No idea could be further from the truth. The horn of plenty, filled to overflowing with every conceivable blessing, awaits but your command to spill its bounty into your life. We have dealt elsewhere with the mental techniques that the sages of the world have always used to bring them what they need in the way of prosperity, and it is not necessary to go over that ground again in this book.

Let us leave you with one final admonition in regard to wealth:

Your greatest material possession is your physical body; As you cherish, care for and use that body, so should you cherish, care for and use all other material possessions which the Creative Power has placed upon your table. And just as the day must inevitably come when you lay aside the body, so on that same day must you release your hold on all material things which the earth plane has given you to ease and make joyful the journey of life. Be grateful that you have smelled the flowers, that you have touched a loving hand, and that Providence has strewn your path with the treasures of the earth.

Chapter 8

Hilarion on How Astrology Works

Most Astrologers believe that the planets, acting through some unseen vibrational influence from their geometric positions at the moment of first breath, in some sense *cause* an individual to have the various characteristics and experiences that are traditionally ascribed to them. For example, there is a consensus among Astrologers that a person born with Uranus on the Ascendant (the eastern horizon) or on the midheaven (the zenith) will literally receive from that planet a kind of overlay or impressed influence which will make him tend to be independent, freedom-loving and unorthodox during his life pattern.

While such Astrologers have correctly identified the traits that are *associated* with Uranus, they have erred in assuming that this planet somehow *causes* the effects to come about. What happens is simply that the guides of an individual (who understand astrological symbolism as well as the best earth Astrologer) *plan* the birth chart for the life about to be lived. To do this, they must search through all possible chart combinations spanning three weeks or so at the end of the pregnancy, in order to find the one which is the best "fit" for the individual concerned.

As an example, suppose that a particular soul has, during recent lifetimes, acquired the habit of taking himself too seriously. Assume as well that his guides expect him to develop the same trait again. In this case, one of the stipulations for the birth cart could well be that at first breath Saturn must be just below the eastern horizon. Astrologers will recognize that this would place Saturn in the "first house" of the chart, thus applying the traits associated with Saturn to the personality or self-image of the person. Now, Saturn is in this position for at least an hour each day, and thus the guides would have a multiplicity of possible charts to choose from. However, they would not base the selection on only this one trait. Suppose that, due to the foreseen troublesome interaction between the individual and his chosen parents, the guides also wanted the birth chart to have the moon in the sign Cancer and in a "difficult" angular relation to Saturn (since this would point to the likelihood that the early home life of the individual would fall short of the ideal in terms of warmth, stability and emotional support).

Now, the moon moves through about 13 degrees of the zodiac each day, and this will pose a restriction for the guides in terms of selecting the date and time of first breath. During the three week span in which the birth must take place, we will assume that the moon does go through Cancer. However, the moon is in this sign for only a little over two days, and during this short period Saturn swings through the first house only twice. The guides will examine both instances to see which is better from a symbolic point of view. Suppose that Saturn is located at 24 degrees of Aries, shifting only a few minutes of arc each day. Suppose further that the moon is located at 12 degrees of Cancer during the first instance, and has moved to 27 degrees of Cancer by the time Saturn again comes to the first house position. Obviously, the second case is the preferred one, since the two bodies are then in square to each other, whereas they form no aspect in the first case.

We emphasize that the above example is simplified to an almost ridiculous extent, and that the guides are able to take into account a great many factors simultaneously thanks to the instruments available to them. However, we think that the discussion just concluded will give an overview of what lies behind the selection of a birth chart.

We would also add that, in regard to the innermost planets (Mercury, Venus, Mars) and also the Ascendant, there *is* a direct influence of sorts that causes an overlay upon the personality about to develop. In other words, the astrological factors do not *all* act on a pre-arranged symbolic basis only. However, in terms of the *timing* of life-events on the basis of astrological transits and progressions, we can state that virtually all such connections are orchestrated by one's guides in such a way that the astrological chart "works". The chart does not *cause* the events; the events are *timed* to coincide with meaningful chart arrangements by transit or progression.

For further elucidation on this subject, the reader is directed to chapter 32 of our earlier book, *The Nature of Reality*.

Section 2
Mars Conjunctions

Chapter 9

Prologue

As an introductory note to this section, it is appropriate to describe the source of the material about to be given, and the method by which it is received.

Due to the efforts of Shirley MacLaine and others, there has recently arisen a new awareness of the concept of "channeling" and the kinds of information that can be attained through that process. The public tends to be sharply divided on the question of what takes place when material is channeled, and whether any such information can be relied upon.

For my part, having "channeled" some 16 books on metaphysical, scientific and occult topics (the Hilarion Series of books), I confess myself persuaded beyond any lingering doubt that channeling is a valid method of obtaining information, and that disembodied sources really do exist that can provide, on demand, a remarkable assortment of answers, explanations and data on a vast array of topics.

It is perhaps unfortunate that the notion of channeling has acquired something of a cult following in recent years. While it is true that a large percentage of individuals can, with a little training, learn how to act as channels for disembodied sources, it is not the case that all such information is equally worthwhile studying. The reasons for this are several. In the first place, a source can transmit through a given individual only that which the person's conscious mind, acting as a kind of filter, will permit. Secondly, the person doing the receiving must have a working knowledge of a particular subject under discussion, otherwise the source cannot pass through the necessary concepts, nor can it find the vocabulary required to express them adequately. Thirdly, not all such sources are at the same level of understanding, nor do they all share the same degree of goodwill toward mankind. Indeed, some of them are clearly of a mischevious or malevolent nature, and are to be avoided.

Consider an example of the first of these limitations. The great 20th century psychic Edgar Cayce, who towards the end of his life was spending up to eight hours a day giving trance readings, was an inveterate coffee drinker. His source stated clearly its view that drinking several cups of coffee each day was beneficial to the system. My contention is that Cayce's consciousness would not have allowed a contrary opinion through, because he himself was addicted to the stuff.

Cayce also provides an example of the second limitation mentioned earlier. Whenever he would get questions regarding Astrology, a subject about which his conscious mind knew only a little, he would answer in generalities. This is not to say that the answers were without value. My point is simply that, had Cayce been a student of the subject, this would very likely have allowed the source to be more specific in answer to the astrological questions. By contrast, Cayce was known to be something of an authority on the Bible. He had read it through many times, and was active in the church for years. As a result, the source which spoke through him was able to quote many scriptural passages accurately, and provided a wealth of background material relating to biblical accounts.

We cannot look to Cayce to exemplify the third kind of problem mentioned earlier. His source was clearly intent on providing benefit to mankind and unquestionably did so for decades. Yet there are sources channeling at the present time which are not so evidently in the category of benevolent entities.

How can one know whether the information channeled from a given source is valid, or that the source itself is trustworthy? Much of the material is unverifiable—like that pertaining to past lives, karma, and so on. For this kind of data, the only test is the "inner voice" of the listener: if the information sits right and "feels" true, then one should accept it at least tentatively. However, there are other kinds of material which can indeed be tested in the arena of reality. Here I speak of scientific and occult information that lends itself to verification.

I realize and accept that the channeled material received through me and published in the Hilarion books should be measured for validity in the same ways as just mentioned. I am, however, encouraged by the fact that virtually all of the information that can be tested has been verified by experiment or observation. The Hilarion Books are filled with data on science, Palmistry, Astrology, symbology, Numerology, psychology, the significance of name-sounds, the meaning of the various teeth, and much more. If the information had failed to be verified, the Hilarion Series would not have generated the interest it now has.

Of particular interest is the theoretical approach to science provided in *The Nature of Reality* and other titles. For example, the use of the inert gases to generate energy fields that can be felt by the hand is something for which conventional science cannot account. Yet this information came through the source Hilarion in a channeled manner. Skeptics who cannot accept the notion of disembodied sources are left with the problem of explaining where the idea could possibly have originated. If from my own "subconscious", then how did it get there? Being an unknown technology, where could my unconscious mind have learned it?

Similar questions must be asked about the information on Palmistry, Astrology, Numerology and general symbology. It was all largely unknown prior to being channeled, and yet it checks out very respectably. There is certainly a likelihood that some of the data was influenced by the limitations imposed by my own personal prejudices, and could have come through in a distorted manner. However that risk is always present, and should not deter the reader from taking what for *him* is valuable in the writings.

In the book, *Astrology Plus,* Hilarion provides a great deal of interesting material on birth chart interpretation. Specifically, he gives a unique approach to interpreting the meanings of the ten astrological indicators (sun, moon, Mercury, etc.) when in any of the twelve zodiac signs and any of the twelve "houses" of the chart. (Readers not familiar with these terms are directed to the introductory material at the beginning of *Astrology Plus*.)

In that first book, however, Hilarion did not deal in depth with the different "aspects" or angles which the various planets can make between each other. One of the most prominent such aspects is the conjunction, which is considered to be present when any two planets in a birth chart are within about 8 degrees of each other. (Some astrologers allow a greater separation for the conjunction.)

The material that follows provides interpretations for the conjunctions of the planet Mars with the other planets, when such conjunctions are found in a birth chart. It is to be underlined that not everyone will have Mars conjoining another planet. Each conjunction is considered by sign. I hope that this information proves useful to practitioners of Astrology, as well as to the general public.

This section of the book also contains several general discussions which will be of interest to all readers, and which are additional to the specific interpretations of Martian conjunctions by sign. I recommend a perusal of Hilarion's Introduction (which follows), and of the preliminary remarks at the beginning of each of the ensuing chapters. Especially interesting are his summaries of the effects of Uranus and Neptune on the development of television and the cinema, respectively.

—M.B. Cooke

Chapter 10

Introduction

The planet Mars has not been fully understood by Astrologers. While its major impetus—that related to energy and its use—has been grasped, it is not as clearly realized that the red planet has functioned as a testing influence for the race of humankind on this earth. Indeed, without the help of Mars, your civilization would have found neither the power to build nor the will to control that power. We wish to speak in greater detail about the role which Mars has played in your history—not only the history of your civilization, but the even greater saga of your coming-of-age as a cosmic sub-race. For the red planet has done much to shape the nature of your collective psyche. By forcing each one of you through the crucible of anger, it has made you aware of the dark seeds that lie within, and of the destruction that inevitably attends any release of fury and rage. It was hoped, by those responsible for placing Mars in your psychic field, that these experiences would teach you that your wild nature must first be subdued before you can take your place among the mature races of the cosmos.

To hold the symbol of this primordial lesson always before your subconscious, you were given one of the noblest of life-forms with which to share the earth: the horse. The power and strength of this animal are known to all, but equally understood is its wild, unruly nature when untamed. Before a horse can be used to carry a rider or pull a plough, it must be "broken", which is the symbolic equivalent of learning to control one's own internal energies so that they can be put to positive use.

Whenever, in an Astrological birth-chart, the planet Mars

is found to be conjunct another of the ten planets within a span of about 8 degrees, you may be certain that the individual's guides (who planned the chart and arranged the timing of the birth in order to produce it) meant to point out something of great importance to the individual. We will be discussing all of the Martian conjunctions in this book, since the particular planet with which Mars is conjoined adds additional information regarding the nature of the characteristic being designated. However, it can be said in a general way that *any* Mars conjunction (regardless of the planet) is a pointer to a problem regarding the control of anger. By using the expression "the control of anger", we are actually implying that there are *two* problems: the presence of the anger itself, and the inability to control it.

Up until recent decades, the history of civilization on this earth has been largely one of war, conquest and plunder. During the time since the Second World War, however, there has emerged a consciousness of the futility of large-scale armed conflict, fuelled principally by fear of the mass destruction that would be unleashed if the superpowers allowed a nuclear exchange to occur. Indeed, it may not be difficult to understand that the guides of this planet deliberately allowed the human group to have access to nuclear power before it was spiritually mature enough to use it safely, precisely in order to foster the heightened consciousness that has occurred.

Of course, not all individuals in the human group have shown evidence of this raising of consciousness. There remains a fraction who are fully committed to selfishness and who see no reason to learn to control their baser instincts. These laggards will be allowed another short space of time to reconstruct themselves along more spiritual lines, but when the time of choosing arrives, no further opportunity to seek higher ground will be granted.

We did not wish to sound like the proverbial Dutch uncle in this book. Indeed, it is not we who determine the timing of the great cycles of civilization on earth. That duty falls to beings much higher on the scale of power and majesty than we. Our task is merely to help humanity to *understand* the patterns through which it has lived, and to perceive the direction in which the currents of its history are inexorably moving.

Let us return now to the matter at hand. While the natal conjunction of Mars with another planet *always* points to a problem with the control of the energies which underlie anger, the nature of the particular conjoined planet allows one to see where, in the individual's make-up, the difficulties are to be found. To grasp the meaning of what we have just said, it is important to realize that an individual does not consist merely of a conscious mind, an unconscious mind and a body, as popular oversimplification tends to assume. For purposes of this discussion, it is appropriate to describe a somewhat more complete picture, since the astrological indicators are capable of rather precisely identifying where in the total pattern the problems are to be found.

We will describe the individual by using the symbolism of the human face. Humanity's guides deliberately established the contours of mankind's physiognomy in order to provide a constant reminder of the true make-up of an individual as he or she manifests on the material plane.

This description will proceed from above downwardly, to reflect the flow of energy from the higher manifestations to the lower.

Consider first the forehead—"the noble brow" of mankind. This part of the face represents the oversoul of the individual. That which is marked on the forehead always bears a direct relationship to the make-up of this ultimate source of one's beingness, and a person properly instructed can read in the forehead lines and other marks a description of the nature of the oversoul—to the extent that the various traits are important to the present life.

At the lower central position on the forehead is the location of the Ajna center, or third eye. This aetheric organ is that with which clairvoyants are enabled to see astral and aetheric forms, and through which sensitive individuals can channel information from higher entities. Symbolically, the third eye on the forehead represents the conduit or "pipeline" which each oversoul has to the God-self, which lies at the heart of every being. It is from this God-self (literally a spark of the Creator) that all life derives.

It is important now to draw a distinction between the oversoul and the "higher self". The higher self (as we shall use the term in this book) is the level directly below the oversoul, where the various characteristics begin to manifest or "crystallize out" in a way that allows them to be assigned to the "male" and "female" categories. On the face, these two aspects of the higher self are represented by the two eyes. The left eye designates the female side, while the right eye denotes the male.

It has oft been stated that "the eyes are the windows of the soul", and the explanation given above will cast further light on the truth of this idea. Indeed, it is a fact that the higher self (often loosely equated with the "soul") literally stands just back of the pupils and does indeed look out of the windows which the eyes provide. Anyone who has had the disquieting experience of looking directly into the eyes of another at the point of death will confirm that something "goes out" of the eyes when the life ends. They glaze over, so-to-speak, and it is obvious that there is no longer any living being behind them.

Reflect for a moment on the striking difference between the eyes and the forehead. The eyes are typically far more arresting, interesting and complex than the forehead, and there is a reason for this distinction. The brow is relatively undifferentiated because at the level of the oversoul the various traits and characteristics which human beings are so used to observing in each other are indistinct and literally "unmanifest". They become identifiable first in the higher self, as represented by the eyes. It is at the level designated by the eyes that the details of human form and function emerge. For example (as iridologists know), the eyes contain within them a microcosm of the physical body, and this is symbologically apt in view of the fact that the higher self literally precipitates the mold for the human body from its own essence.

The eyes are also associated with other facial features that have symbolic meaning: the eyebrows and the upper and lower eyelids. Although we are writing a further full-length book to deal with all facial features and their meanings, it will suffice here to make two brief points. First, the lids represent the protective influences which the higher self plans for itself during each incarnation: the upper lids denoting the material level of protection, the lower lids pointing to the *psychological* protection which is afforded by the parents and their symbolic archetypes in the subconscious. Second, the prominence or otherwise of the eyebrows symbolizes the degree to which the oversoul is attracted toward manifestation on the earth: bushyness usually translating into a heightened sensuality which makes the material plane seem more attractive. Note the position of the eyebrows along the *bottom* of the forehead, symbolizing the idea that the prospect of enjoying sensual or physical pleasure has been responsible, at least in part, for attracting the oversoul "downwardly" into incarnation.

We come now to one of the most prominent features of the face: the nose. This invariably represents the ego or self-picture. The shape, colour, size and markings of the nose can be read like a book by one properly instructed, although we will leave these details for elaboration in our book dealing specifically with the human face.*

Note that the nose emerges from between the eyes and projects downward toward the mouth. This represents the idea that the ego is generated from the interplay of the male and female facets at the level of the higher self. In order to make a new vehicle (human body) for itself, the higher self literally extends a kind of probe downward into the lower or material spheres during the gestation period, and this acts as a kind of core or central mold about which the other facets of the earth personality are precipitated and gathered. In essence, this probe is the ego.

The nose is located at about the middle of the face, thus standing mid-way between the higher manifestations (represented by the eyes and forehead) and the lower or personality manifestations. The latter are denoted by the upper lip, the mouth and the chin.

---

* This book is entitled *Faces*.—Ed.

The symbology of the face, as designed by the guides of humanity, recognizes three distinct facets of the lower self or personality: the mental, the emotional and the physical. The first two of these are represented on the upper lip, while the physical body is shown on the area from the mouth down to the chin. Dealing first with the upper lip, this facial feature extends from the bottom of the nose to the mouth. The upper region close to the nose is connected with the mental sphere, while the lower extent just above the mouth denotes the emotions. The symbology here is very apt, since the bottom of the nose is a passage for air (which represents the mental aspect) while the mouth is normally always wet with saliva (wetness and water denoting the emotions). The symbology extends even further: the nostrils at the bottom of the nose represent the subconscious, compartmentalized into the left side for female aspects and the right side for male aspects.

The relative sizes of these lower regions of the face are in proportion to the prominence or degree of importance given to the various facets being represented. For example, large nostrils represent a prominent subconscious containing a great deal of information and many response patterns which impinge importantly upon the daily life of the individual. Likewise, a prominent chin and strong jaw line denote a physically oriented individual, i.e. one who tends to place more stress on the physical than on the emotional or mental. There is no need to describe this correspondence further, as we will be dealing with every facet in detail in our separate book on this subject, including how to read the timing of life-events by the specific positioning of marks, scars, etc. on the various indicators we have mentioned.

It will now be recogized that the "map" of the human face provides an excellent chart by which to perceive the flow of formative energies that have made the individual what he or she is. To summarize, the face bears representations of the following facets of the total individual:

— the God-spark
— the oversoul
— the higher self (male/female facets)

- the ego
- the subconscious
- the personality, including
  - the mental facet
  - the emotional facet
  - the physical facet

We are now ready to delve into the meanings of the various Martian conjunctions as these may appear in a birth chart. The student of Astrology is asked to keep in mind as well that whenever progressed Mars (one ephemeris day = one year) conjuncts a natal planet within one degree, or vice versa, there arises a time window within which an important test of one's ability to control or deal with anger can be encountered.

The chapters which follow address each Martian conjunction in order, beginning with the sun.

Chapter 11

Mars-Sun Conjunctions by Sign

We will deal first with the solar conjunctions in each of the twelve signs.

We wish to stress an important point with regard to the planet Mars. *Any* conjunction of Mars with another planet in a birth chart is *always* to be regarded as an affliction to the other planet, although it may not be an affliction to Mars. For example, Mars conjunct the sun always debilitates the sun, and tends to bring out the more negative or difficult aspects of the sign and house in which the sun is located at first breath. However, this same conjunction is not an affliction to Mars. The only conjunctions which tend to afflict the red planet are those of Saturn and Neptune: Saturn because the ringed planet *always* afflicts any other planet with which it is in aspect, and Neptune because it is so discordant with the energies of Mars, tending to sap the strength of the Martian placement. It will be remembered that the latter conjunction shows up more prominently than chance expectation in the alcoholism study set out earlier in the present book.

The sun designates that aspect of the oversoul which has been tapped to generate the "higher self" in the present life-incarnation. It is important to realise that not all facets of the oversoul can manifest in a single life, for most individuals. Usually, there are simply too many unresolved conflicts between the different past life experiences and acquired habits to allow all of the oversoul to make its presence felt. For this reason, a selection has to be made, and the sun position by sign represents symbolically some aspect of the selected qualities.

Hence, we can say in a general way that the sun in the natal chart symbolizes that part of the "soul" which is manifesting in the present life. Our book *Astrology Plus* sets out in detail some of the aspects or complexes which the sun can represent by astrological sign.

However, when the sun is natally conjoined by Mars (within 8 degrees), one of the meanings is that there are unresolved energies of anger that are as yet firmly rooted in the oversoul, and that serious effort must be put forward in the present life to transmute the fury and rage at this highest level into something more positive. The way to go about this is to follow the "flow-path" represented by the face, and to draw the tensions down from one stage to the next until they become available at the level of the personality—where they can more easily be dealt with.

The present book is not an appropriate place to deal with this therapeutic approach in detail.

Let us instead turn now to the Mars-sun conjunction in the different zodiac signs, to see how the astrological filters clarify further the nature of unresolved anger at the oversoul level.

Aries

In Aries, Mars is essentially well-situated. The red planet is regarded as the ruler of this fiery, impetuous, physical sign, and the placement of the conjunction here indicates that the unresolved anger, though present, is able to be channeled into positive avenues. Especially, this individual will have a great capacity for athletic endeavor, provided the strength of the body is sufficient. He or she should, however, strive to keep the energies at a simple competitive level, and avoid becoming personally committed to winning. The traditional self-absorption of the Aries presents the danger of coupling the self-image to the scoreboard. There is a risk that losing will release the negative side of the energy, and anger, self-scolding and resentment will result. It is for this individual that the old maxim was penned: "It matters not whether you won or lost, but how you played the game."

Taurus

Here, the conjunction draws its power from the deep well of unyielding, immovable rootedness that is so characteristic of this earthy, fixed sign. Mars is debilitated in Taurus, being opposite Scorpio where it is one of the rulers (along with Pluto). Here, in its fall, the red planet must struggle particularly hard to show any positive effect. The sign Taurus, ruled by the gentle and loving Venus, colors the blocked conjunction energies with the hues of uncontrolled passion, and there is a danger that these passions will sweep away the more tender side of the male-female relationship in a sexual tide. The fixed nature of the sign of the bull makes any correction or softening of this approach difficult to achieve. What is happening is that the unresolved energies of anger at the oversoul level are being vented in reckless passion, the hope being that sorrows arising from this behaviour will prompt a moderation of the tendency, which by a parallel effect will begin to correct the imbalances at the upper levels.

Gemini

In this airy, surface-skimming, double-bodied sign, the Mars-sun conjunction tends to produce a mental aggressiveness that is easy to spot. In addition to bringing to the fore the basic Gemini lessons of sarcasm and scattered forces (due to the afflicted Sun), the conjunction is indicating that the unresolved energies of anger at the oversoul level are finding an outlet in the mental sphere of activity. The individual with this placement should seek to use this power for personal achievement rather than to wound or discomfit other people. There is a strong tendency to fire the energy off in all directions, and little determination to channel it into any long-term endeavor.

Cancer

In this watery, emotional, self-absorbed sign, the Mars-sun conjunction represents a difficult problem to be overcome. The problem is essentially that the ego is fragile and cannot allow slights (whether imagined or real) to pass unchallenged.

Unfortunately, many of the presumed slights are indeed imagined, arising from the fear orientation of the Cancer personality. Mars in this sign is in its fall (opposite Capricorn where it is exalted), and this placement exaggerates the intrinsic mistrust with which the Cancerian approaches life. The energies of anger at the oversoul level are almost all assimilated into this fearfulness governing the life pattern, which fuels the fury with which the Cancerian attacks those who are labeled enemies.

Leo

Here, in the expansive, regal, blazing sign of fixed fire, Mars is not badly placed—nor is the solar conjunction, as the sun rules Leo. The primary outlet for the high-level energies of anger will be disdain and haughtiness toward others, and vindictiveness when crossed. Added to this will be the exaggerated pride of the afflicted sun, and the inability to admit mistakes or take blame.

Virgo

In this mutable, detail-oriented, critical sector of the zodiac, Mars has some difficulty in expressing its power. By afflicting the sun, the red planet produces an increased tendency to be idealistic and critical of others. If Mars is itself afflicted (the solar conjunction is not an affliction), then you may be certain that the unresolved oversoul energies will make themselves felt in the personal relationships. These will tend to be marked by brief infatuation or idolization, followed by crushing disappointment and anger when the other person turns out not to be a god or goddess after all. Even if it is essentially unafflicted, the Mars-sun combination will lead to disillusionment in the personal life, primarily because the sights are habitually set too high. Such individuals must realize that the world is an imperfect place, full of imperfect people. If he or she can learn to cherish the foibles and idiosyncracies of others, rather than to use them as an excuse for criticism, life will become much less disappointing.

Libra

Librans are generally a pleasant lot, eager to set others at ease, willing to take pains to promote smooth interpersonal exchanges. However, when Mars conjoins the sun in a Libran chart, the message is one of caution: the unresolved energies of anger at the oversoul level are likely to burst forth into the personal relationships in a disruptive way. The red planet is in its detriment in Libra (being opposite to Aries, which it rules), and this alone would suffice to block some of its power. By conjoining the sun, Mars points to a tendency to "be one's own worst enemy" in terms of love-life. Typically, after much effort has been expended to establish a relationship on a firm footing, the Mars-afflicted Libran will suddenly and inexplicably ruin everything in a single outburst of rage, accusation or jealousy. This is the oversoul conflict manifesting. The hope is, of course, that in the aftermath of such destructive behavior, the individual will try to understand and correct the internal flaw that caused the heartbreak. By doing so, he or she will simultaneously be performing a similar corrective therapy at the level of the oversoul.

Scorpio

Mars is powerfully situated in this brooding, intense, manipulative and emotional sign. By conjoining the sun here, with the Solar body ruling a sign which is in afflictive aspect to Scorpio, the message is that the individual must learn to bridle the Martian horse. All too often, the scorpionic Mars throws itself into sensuality, seeking in the physical passions an escape from the sense of futility and impending doom which seems to follow one around like some dark, vindictive specter. Even more important, however, is the magnification of the scorpionic trait of all-out retaliation against those who are conceived to be enemies. The individual must learn to fight fairly, to make allowances, and finally . . . to forgive.

Sagittarius

This placement of the Mars-sun conjunction heightens the

enthusiasm and childlike glee with which the individual approaches the world. The primary problem is that of wishing to remain a child throughout the incarnation. In its extreme form, this trait can undermine the structure of the adult life, leading others to avoid trusting the individual with responsibilities and important projects. Often the unresolved anger is channeled into athletic or other physical activities, and turns to sourness and resentment only in later years when the Mars-afflicted Sagittarian realizes that he has scattered his seeds on barren land, and foolishly squandered the great gift of life.

Capricorn

In this Saturn-ruled sign, Mars is well placed (its exaltation). Thus, even though the sun is considered to be afflicted by the closeness of the red planet, nonetheless there are distinct possibilities of making much forward progress in this life, by utilizing the energization of the combination to the full. Either planet on its own in Capricorn would instill a drive to accomplish something, to "make one's mark". Together, they apply even more pressure in this same direction. What is happening is that the unresolved energies of anger at the oversoul level are being channeled down into the material plane, impelling the individual to direct them into practical and worldly endeavors. Here is a person who can pour his whole being into the pursuit of power, money or fame. His primary lesson is not to neglect the gentler side of his nature in the process, nor sacrifice human warmth on the altar of success.

Aquarius

The conjunction of Mars and the sun in Aquarius points to a considerable honing of the intellectual faculties. Cleverness and mental agility usually mark anyone who has been born under this pattern. A major lesson is to remain open to affectional interchange. There is a tendency to assume that the mental sphere is the only arena in which life should be carried on, and that feelings and emotions are merely bothersome impediments that get in the way of the intellectual clarity which the individual prizes.

Pisces

When the sun is conjoined by Mars in this watery, feeling-oriented sector of the zodiac, the unresolved conflicts at the oversoul level are expressed through the emotions. Anyone born under this pattern should seek always to view life in the clearest possible light, to avoid self-pity and to project to others an aura of empathetic concern. There is a great danger that the buried conflict will simply fuel resentment and morose self-absorption ("Why me? What did I do to deserve this?" etc.). Pisces is ever prepared to play the role of the hapless victim, and this tendency derives additional power when Mars conjoins the sun.

Chapter 12

Mars-Moon Conjunctions by Sign

Conjunctions of the red planet with the moon point clearly to the ego or self-picture as the location where unresolved conflict has become lodged. This is represented on the face by the nose. Martian conjunctions with luna also present a picture of confusion with regard to the male-female polarity. The moon is clearly a female significator, while the energies of Mars are linked with the male side. Because of this, it may be possible to detect in the eyes themselves some suggestion that the one principle has been confounded with the other. As an example, look for eyes that are closer together than the average distance, or a tendency for one iris to be displaced from the middle position (e.g. more white on one side of the iris than on the other. The nose itself may also indicate by its shape or position that the conflict is working itself out in terms of the male-female interaction, in concert with the ego or self-picture. For example, a nose that is clearly displaced from the precise midline of the face can indicate a lopsided approach to the masculine/feminine equation. Such a person may tend to favor one side against the other, or may have an inordinate fear of "slipping" over the demarcation line between the sexual roles. A woman with a right-leaning nose may experience too strong a pull from her own masculine tendencies, and out of fear may try to overcompensate by wearing particularly feminine clothes, or by avoiding fashions which emphasize maleness (e.g. shoulderpads), or by playing the role of the "clinging vine" or "dumb blonde" whenever men are present.

The meanings of the various Mars-moon conjunctions are as follows:

Aries

Mars rules this powerful sign, and is thus well placed. By contrast the fluid, emotional moon is out of its element in Aries. Hence the major female significator is under double attack. This signifies that the feminine principle, as it is found in the self-picture, is distorted or weak in some way. A man with this pattern will often make a habit of choosing women for all the wrong reasons, and his relationships will tend to be stormy and exhausting. He must learn in this life to see the female archetype clearly, and if he can do this he will discover that males and females are not as different as he grew up believing. The clue to this truth is found in the "normal" face, in which the two eyes are basically identical, though mirror-reversed from one another. When a woman has this conjunction, there is usually a tendency to feel inadequate or uncomfortable in the traditional female role. This arises because, unless great effort is put forth, the archetype of feminity cannot be separated from the aggressive energies that are locked up in the ego.

From the level where this dictation is originating, the ancient rivalries, suspicions and resentments between the sexes are viewed as a major stumbling block that will have to be cleared out of the path before the human race can walk with trust and joy into the light of the coming New Age. Men have traditionally kept women in a subservient role, often out of well-veiled feelings of insecurity. And women have in many cases been willing accomplices to the travesty. The only path around this blockage is for individuals to realize that, apart from the biological roles, there is little difference between the sexes. The solution is not to replace patriarchy with matriarchy, but for each side to acknowledge the integrity and freedom of the other. We expect that this balancing of the equation will occur automatically, as more and more individuals realize that they have incarnated many times in both sexes, and that the current sexual identity is merely a mask which the oversoul has donned for the purpose of learning something important.

Taurus

When luna conjoins the red planet in this earthy, pleasure-oriented sign, the individual has a strong inclination toward self-indulgence in the sexual area. The confusion between the male/female energies gives rise to an approach which seeks to underscore the individual's sexual identity through this kind of excess. Again, the conflict is lodged or 'stuck' at the level of the ego or self-picture. Look for a swelling or enlargement of the lower tip of the nose. Its presence will indicate that the individual is still working on this problem. Its absence suggests that at least some progress has been made.

Gemini

The conjunction of Mars with the moon in this airy, intellectually oriented sign points to the mental arena as that in which the imperfections or exaggerations of the self-picture will certainly manifest. The individual will tend to be a "mental warrior", ready to do battle with anyone so long as the weapons are the rapiers of wit, sarcasm and caustic utterance. The best palliative for this condition is a sincere attempt to allow others to be less mentally acute than oneself, without relegating them to some brutish, underserving sub-race. Remember that no amount of cleverness can ever justify an unkind word, nor can bare intellect alone discover the truth of the heart. Typically the arms or hands will be subject to accidents or below par in some way.

Cancer

When the red planet conjoins luna in this moon-ruled, watery and self-absorbed sector of the zodiac, the main indication is that of excessive caution and the mistrust of others' motives. Here is a person ever ready to take remarks the wrong way, to assume a slight when none is intended. The cure for this condition is to learn *trust*—not only a trust in other people, but a reliance upon the benevolence of Providence. Guard against stomach conditions.

Leo

In the proud, noble, fiery sign of the cosmic lion, Mars is in a compatible vibration. Not so the moon, however. The result is a debilitation of the female archetype in the psyche. In many cases, the individual will be able to point to a younger portion of the life-experience when the person most clearly representing the feminine principle was absent, ill, or in some other circumstance which limited the interplay with the owner of this conjunction. Often this means the mother, but there are many cases where a sister, aunt or other relative plays this role. The most important thing here is simply to *recognize* that this pattern has affected the development of the ego or self-picture, so that the more serious negative results can be moderated. (This conjunction also points to a youthful phase when the self-picture was damaged by a personal failure of some kind.)

Virgo

Neither Mars nor the moon is particularly comfortable in this mentally-oriented, practical and picky sector. The conjunction of these two bodies in Virgo will tend to bring out all of the critical capacity of this person, and in extreme cases he or she will drive others away by the relentless fault finding. The antidote for this unfortunate habit is simply a realization that one has this tendency. (The conjunction here also suggests a time, in youth, when the individual was caused embarrassment by an older person's misadventure, failure or bankruptcy.)

Libra

Mars is in its detriment in this emotionally controlled, cerebral and sociable zodiac sign. The moon fits awkwardly as well, and therefore the combination of the two will produce a general feeling of being a square peg in a round hole. This is the mark of the "outcast", the person who feels shut off by others. In reality, it is not others but the *self* that has closed the door. And only the self can open it. The most important thing here is to soften the conviction that one does not *need* other people, for this is the thought-pattern that has driven them all away. (The Libran conjunction of Mars and the moon also tends to exaggerate a tendency to be indecisive.)

Scorpio

The moon finds itself under double attack in this pattern: she is in her "fall" (being in the sign opposite the lunar exaltation), and is conjoined by the red planet (which afflicts by conjunction). It is therefore almost certain that some unfortunate event in the early life has warped the formation of the feminine principle in the psyche. Often, though not always, the event had some sexual overtones. In most cases, the individual will readily recollect the circumstance to which this conjunction points. As in many things, knowledge and understanding can allow the negative results of the episode to be overcome. However, the individual will usually need to undertake a serious program of counteraction in order to be delivered from the thought-patterns that were created around that early incident.

Sagittarius

This Jupiterian sign is an anomaly in the zodiac, in that the beneficence of the Jovian influence uplifts almost any planet found in the sector. For this reason, a Mars-moon conjunction here does not carry the same heaviness that it brings elsewhere. The primary meaning of the combination is that of an excess of playfulness, enthusiasm and/or naïveté. The self-picture is likely to be colored by youthfulness and much physical activity.

Capricorn

Here Mars is in its exaltation, the moon in its detriment. The debilitation of the lunar body points to an early home life circumstance in which the individual failed to acquire a balanced assortment of archetypes. Typically, the early portion of the incarnation will be characterised by an austere, controlling influence, with little spontaneity and even less laughter. A major task in the life pattern is to rise above the limitations which the dismal aspects of the early life have imposed. Guard against knee trouble during youth.

Aquarius

The conjunction of luna with the red planet in this intellectual, emotionally controlled sector typically corresponds with considerable mental gifts along with a tendency to approach everything from the vantage point of the mind. Such individuals normally will not recognize that they have this trait, so accustomed have they become to their way of viewing reality. Unfortunately for them, this attitude removes from their grasp the soaring emotional experiences that only love and affection can engender. Guard against circulatory disorders.

Pisces

The moon is relatively well placed in this watery, sentimental and indecisive sign, but Mars is not. The conjunction here points usually to a life-pattern in which the self is paraded as a victim of circumstances over which the individual pretends to have no control. In actual fact, of course, each person projects his own reality and carries full responsibility for that reality. The best advice one can give is to stop wallowing in self-doubt or self-pity, and instead to seize the reins of life and take charge of one's future. Unfortunately the owner of this conjunction will often find this a Herculean task and will tend to lapse into an even deeper abyss of gloom when he or she perceives a lack of progress. (This conjunction also can point to trouble with the feet.)

Chapter 13

Mars-Mercury Conjunctions by Sign

The conjunctions of Mars with Mercury generally indicate that the unresolved conflict represented by Mars is manifesting in the lower personality, specifically in the mental sphere as represented by the upper lip just beneath the nose.

Those who have this conjunction in the natal chart will often (though not always) have some mark, scar or other indicator just below or beside the bottom of the nose.

Generally, regardless of zodiac sign, this pattern identifies one who knows well the power of words to wound others. A major task in this life is to learn to control what one says.

Here are the ancillary meanings for the various signs:

Aries

With Mars in its home sign afflicting Mercury by conjunction, there will be a tendency to be verbally assertive and testy toward others. Expect that one of the parents served as a model for this trait.

Taurus

Mars is in its detriment in this sign, and Mercury is also less than ideally placed. One major indication here is that the neck and/or head are weak in some way. Guard against throat infections (a karmic consequence for having verbally attacked others in previous incarnations).

Gemini

The conjunction of Mercury in its home sign with the energetic Mars will tend to produce a powerhouse of verbal abusiveness toward those conceived to be enemies. If this trait has been allowed to run unchecked, the individual will find that he has estranged most of the important people in his life.

Cancer

The conjoining of Mars with Mercury in this watery, feeling-oriented sign often produces emotional immaturity, and a tendency to allow the emotions to influence the thought-processes unduly. The unresolved anger often surfaces in childish outbursts or tantrums.

Leo

In this prideful sector, the Mars-Mercury conjunction often exaggerates a tendency to view the whole world as playing second-fiddle to oneself. Self-absorption and false pride can be the downfall of the person with this placement. The unresolved energies of anger typically emerge as haughtiness.

Virgo

Mercury is well placed in this practical, earthy sign, but the flow of Martian energy is blocked by the excessively analytical nature. Look for some physical weakness, or a tendency to become drained of energy when opposed. The main problem, however, is the excessively critical trait, due to the affliction to Mercury.

Libra

The conjoining of Mercury with Mars in this controlled sign will cause a person to avoid any display of emotion—whether positive or negative. This rigid stance, unless softened during the life pattern, will tend to estrange others and produce the feeling of being an outcast.

Scorpio

In this brooding, powerful sector, the Mars-Mercury conjunction emphasizes the sexuality of the individual. Unless internal corrective steps are taken, sex becomes simply too much of an issue. Not infrequently, the person with this placement undergoes the loss of someone close, typically someone younger. This trauma is intended to remind the individual that there are many things in life that are more important than sexuality for its own sake.

Sagittarius

In the sign of the Archer, Mars conjoined with Mercury produces a pronounced youthfulness verging on immaturity—to such an extent that it gets in the way of other departments of the life. The individual with this placement needs to be reminded that one can be responsible and enthusiastic at the same time.

Capricorn

Mars is exalted in this Saturnian sign, but Mercury has difficulty expressing itself here. Together, they paint a picture of one who can funnel much mental energy into the career, but who tends to be of too serious a cast of mind. The individual needs to "let his hair down" from time to time.

Aquarius

Mercury is excellently placed in this airy, mentally oriented sign, producing a keen and broad-ranging intellect. The conjunction with Mars adds extra mental energy, and even though it is an affliction it cannot take away the sharp edge of the mind. However, one can expect the Martian influence to draw one's thoughts toward sexual fantasizing, and thus to waste some of the mental power.

Pisces

Neither Mars nor Mercury is well-placed in this watery, emotional sector of the zodiac. In conjunction, with Mercury getting the worst of the combination (being in its detriment), the indications are that the individual will fritter away much mental energy in self-pity, in dwelling on dark thoughts, in going over and over the scenes of past sorrows, and in trying to comprehend that which cannot be comprehended from an intellectual standpoint alone. Help for such an individual can only come through an internal house-cleaning, in which he decides which subjects deserve his mental energy and which do not.

Chapter 14

Mars-Venus Conjunctions by Sign

The conjunctions of the red planet with Venus generally indicate that the unresolved energies of anger have become lodged in the emotional facet of the lower personality. On the face, this region is identified on the upper lip, just above the pink. One who has this conjunction natally will often show one or more marks, scars or the like on this area.

As a general comment, this conjunction—regardless of the sign in which it is found—designates a tendency for the affectional relationships to be the arena in which the unresolved anger seeks its resolution. The hope is that the emotional bonds which the individual forges will become important enough to encourage him to work hard at undoing the knots of conflict with which he was born. As a result, until he succeeds at this task, his love involvements will tend to be characterized by storminess, conflict and disappointment.

Aries

Here Mars is well placed by being in its home sign and conjoined by Venus (considered a positive aspect), while Venus is badly placed in its detriment and afflicted by the Martian conjunction. As a result the combination will provide lots of energy but a star-crossed love life. Infatuation and disillusionment will come again and again until the individual learns this important lesson: that love needs no physical connection to shore it up, but exists—perfect and complete—at its own separate level of being.

Taurus

In this sign Mars is afflicted by position, while Venus is well placed (being the ruler of Taurus). The basic meaning is that love-affairs for this individual will loom large in importance, even to the extent where they are overly emphasized. Until he can put this facet of experience in proper perspective, the affectional experiences will continue to be painful.

Gemini

Neither Mars nor Venus is comfortable in this mentally oriented sign. The conjunction can be expected to produce a "skim-the-surface" attitude toward commitment in terms of the love relationships. The breezy approach to affectional experience will typically lead to sorrow until the individual learns to take promises more seriously.

Cancer

Mars is badly placed in this watery, emotional sign, while Venus is afflicted by the Martian conjunction. Love relationships will tend to be characterized by excessive emotionalism, insecurity and suspicion. The most important lesson suggested by this pattern is to learn to trust one's partner.

Leo

This person wants to have the "perfect" love, the ideal relationship in which the shining knight on a white horse carries the damsel off to a perpetually happy ending. Unfortunately, life is not like that, and until the individual lowers his sights to something of a more realistic nature, he will be disappointed again and again.

Virgo

In this sign of practicality and nit-picking, the Mars-Venus conjunction points to one who will not settle for anything less than perfection in the love-partner. (This is distinguished from the Leo entry above, where it is the *story* that must be perfect, rather than the actual mate.) This individual must realise that everyone on the earth plane is flawed, including himself, and that it takes a well-developed, mature affectional nature to love someone *despite* his failings.

Libra

This sector of partnerships is ruled by Venus, but is opposite the sign ruled by Mars (Aries). As a result, Venus is relatively well placed (though afflicted by the Martian conjunction), while Mars is debilitated (though aided by the Venusian conjunction). One with this pattern in the natal chart will typically have an ambivalent attitude toward the idea of the pair-bond, finding it both attractive and repugnant at the same time. Relationships will tend to be marked by a lurking suspicion that the partner doesn't really feel the same way, or is just pretending. Thus, the person with this placement will often feel isolated or cut off even when in an affectional situation.

Scorpio

In this sign the two conjoining planets are clearly polarised: Mars benefitting by sign and by its conjunction with Venus; the latter debilitated by sign and the conjunction with Mars. Generally, one can assume that the life pattern will be marked by at least some liaisons in which "love is destroyed by sex". The individual must learn that physical passion is intended to be the *servant* of love, not its master.

Sagittarius

In this Jupiter-ruled sign, where afflicted planets often point to some form of exaggeration, the positioning of Venus afflicted by the conjunction to Mars will normally mean that the contest between the love-nature on the one hand and adventuresomeness on the other will be a major issue in the life-pattern. This is usually brought home by requiring the individual to carry on a love-relationship at long distance for a period of time (as we have pointed out in our book, *Astrology Plus*). The well-aspected Mars in Sagittarius (thanks to the Venusian conjunction) points to a youthful approach to most things throughout life, and an abundance of energy for personal projects (but not for drudgery).

Capricorn

Here we find Mars superbly positioned—in its exaltation and conjunct what is called the *lesser benefic* (Venus) in traditional Astrology. Venus on the other hand is crushed by the coldness of Saturn-ruled Capricorn, and attacked by the *lesser malefic* (Mars). The combination will tend to produce plenty of energy to funnel into worldly goals (the career), but a rather dismal showing in terms of the love-life. The individual would be well advised to value human warmth more (and to *show* it to others!).

Aquarius

The conjunction in this sign of mental clarity, independence and unorthodoxy tends to produce a person who lives his own version of life, regardless of what others think. This incarnation will allow him to "break free" of old set ways of doing things, especially in terms of the human interrelationships. A mental powerhouse, but some tendency to dwell on sexual fantasies. The badly afflicted Venus designates emotional coldness or over-control.

Pisces

Sentimentality gets in the way of true affection, with this pattern in the natal chart. Venus is afflicted by the Mars contact, and tends to produce an excessive emotionalism that leads to sorrow in love. Then, when things go wrong, the person asks, "Why me? What did I do?" The remedy is to see oneself clearly, which is easier said than done for anyone with strong Piscean planets. Nonetheless, with effort this can be accomplished. The physical energies can become drained whenever this individual indulges in excesses of any kind, due to the positioning of Mars. The positive contact from Venus, however, allows quick recovery through rest and moderation.

Chapter 15

Mars-Jupiter Conjunctions by Sign

Regardless of the sign in which it occurs, this conjunction should be looked on as being primarily positive in meaning. Mars itself receives a very powerful benefic influence from being conjoined by Jupiter. Hence, one can expect a robust physical body (so long as other indicators do not run contrary), lots of energy, and a healthy sex drive. Jupiter is admittedly in affliction due to its contact with the *lesser malefic* (Mars), however even when badly aspected, Jupiter usually denotes no worse a trait than a tendency to exaggerate various aspects of experience.

This conjunction is an exception to the rule enunciated at the beginning of this section, in that it does not, by itself, indicate unresolved anger. Indeed, it suggests that in previous incarnations such locked-up energies were, through effort, released in a positive way, and that they now provide the power that drives the physical vehicle.

Aries

Here Mars is doubly blessed, for it is also in its home sign. Look for plenty of energy in this individual, but warn him against a lack of caution. The physical body is not indestructible, regardless of how invincible he feels.

Taurus

There is a tendency to exaggerate the importance of material things, due to the affliction to Jupiter. Also, guard against physical self-indulgence.

Gemini

The conjunction in this sign is quite positive in terms of the mental area. The mind is powerful, and the outlook is broad and benevolent, accepting even that which is not immediately understood or believed. Avoid over-emphasis on the importance of the mental sphere, and seek to balance it by encouraging a richer emotional life.

Cancer

Jupiter is content in this sign (its exaltation), but Mars is in its fall and thus rather out-of-sorts. Expect an alternation, during the early home life, between storminess (Mars) and tranquillity (Jupiter). This may be due to the individual himself, who with his birth brings into a relatively harmonious situation the unresolved conflicts represented by Mars. Expect as well, however, an important figure in that early home life who will be instrumental in helping the individual to overcome this disruptive internal trait.

Leo

The main indication of the conjunction in this regal sign is the risk of an exaggerated sense of arrogance and pride. This derives from the affliction to Jupiter. The positively aspected Mars tends to produce mainly pomposity.

Virgo

Neither of these two planets is well placed in this earthy, mentally oriented sign. The conjunction, particularly if within 3 degrees of exact, primarily indicates a prickly conceit which delights at finding fault with others. There tends to be an exaggerated view of one's own mental prowess.

Libra

This conjunction, occurring within the Venus-ruled sign of sociability and harmony, points to one special lesson that must be learned in this incarnation: it is that of coming to terms with *oneself*. In previous experiences on the earth plane, the individual was unable to accept certain of his own traits or characteristics, typically denying or disowning them. However such an attitude produces a splitting apart within the self, and a lack of wholeness which usually leads to physical or mental illness. If one is unhappy with certain of his qualities, the solution is not to deny them, but to change them.

Scorpio

When Mars conjoins Jupiter in this sexually oriented sector of the zodiac, the chart is giving a signal and a warning: the individual must learn to see sexuality as more than merely a means to propagate the race. It was intended to be the crowning glory of true love between mated souls, and a brief taste of the bliss that awaits those who, overcoming the world's trials, graduate to higher levels of experience. If one persists in the mechanical approach to this supreme mystery, then in future incarnations the joy of the experience will gradually be diminished until, like a candle that has burned too low, the flame at last will die.

Sagittarius

The conjunction of Mars and Jupiter in the sign that is ruled by the latter points to an interesting condition: over several previous lives, the individual has brought with him into incarnation too much in the way of talents and abilities. This is not to say that one should not be capable and talented. But in the case of this pattern, the manifestation of multiple aptitudes and gifts has led to a certain laziness, and an assumption that one can do anything as well as or better than others. In the present life-pattern, however, there will be a requirement to work harder in order to develop the gifts that were there before, so that the individual can once again learn to be grateful for his talents.

Capricorn

In this Saturn-ruled sign, the conjunction in question denotes a tendency to assume that success in the world's eyes is the only goal worth pursuing. The individual has a great capacity for work, and a drive to succeed. However, there tends to be a lack of concern for the feelings and wishes of others, particularly if these are seen as impediments to the success that is desired.

Aquarius

When Mars conjoins Jupiter in this intellectual sign, the individual has a tendency to assume that others have the same outlook on life as he, in particular that they regard the mental gifts as highly as he does. It is difficult for one with this conjunction to realise that many people do not revere the power of sheer intellect to the same extent as he, but instead cherish human warmth and affection—or a cosy home life, or the pleasures of eating and drinking—far ahead of the more austere joys of the mind.

Pisces

The conjunction under discussion, when it falls in Pisces, generally designates one who tends to feel that other people are more fortunate, or have a better marriage, or are simply *happier* than he is. On closer inspection, however, it would be found that this individual is no less blessed than others. The important thing is to *realize* that this is so.

Chapter 16

Mars-Saturn Conjunctions by Sign

The essential key idea for this conjunction is that of *yoked power*. In this life-pattern, the energy represented by Mars is intended to be harnessed to some serious purpose. Typically, in earlier incarnations the power available to the individual was either squandered fruitlessly or made to serve ignoble ends. Now, however, he must learn to take command of his own energies, and direct them toward some worthy goal.

Aries

In this sign, the Mars-Saturn combination suggests that personal growth and internal reconciliation will provide the arena in which the Martian energies will be tested and trained. This individual should understand that many trials will come in life, all of them being opportunities to harness the internal power to the task of self-improvement.

Taurus

Here, the conjunction in question points to the love-nature as a likely area where the energies of Mars need to be tamed, trained and controlled. The temptation to indulge in various forms of pleasure is a strong factor in the daily life, and while pleasure *per se* is not a negative thing, nonetheless it is important to distinguish between those indulgences that serve higher goals and those which merely debilitate the mental and physical bodies.

Gemini

In this Mercurial sign, the conjoining of Mars with Saturn designates the mental sphere as that in which the impetuous energies of the red planet need to be controlled and directed. As well, the afflicted Saturn suggests a tendency toward depression or excessive seriousness in the mind. Instead of giving himself up to verbal sparring and abusiveness, this individual should yoke the Martian power to the task of overcoming whatever of despondency and gloom can be found in his nature.

Cancer

Neither of the conjoining planets is happy in this distrusting and emotional sign. With Saturn in its detriment and Mars in its fall, expect that the early home life will reflect some of the difficulties with which these planets are connected. Storminess and sorrow (especially in relation to men) are likely to have been evident in the early years of this individual's life-pattern. The task before him is to learn to master his internal energies, and to take responsibility for his own part in the troubles that have afflicted him.

Leo

When Mars (the *Lesser Malefic*) and Saturn (the *Greater Malefic*) are together in this sign, it is certain that the lesson of *pride* must be worked on in the present life-pattern. In addition, there is generally a tendency to be uncaring about those perceived as being of lesser station in life. This incarnation presents a great opportunity to learn to be kind to those less fortunate, because one of the parents is (or will have been) "inferior" in some way to the individual who has the Mars-Saturn conjunction. This could be in terms of talent, intellect, position, wealth, etc.

Virgo

The conjunction under discussion, when it occurs in Virgo, points clearly to the need not only to curb critical remarks, but to overcome criticism *per se* (i.e. not even to allow critical thoughts to surface). For someone with this conjunction, success with this major lesson may not come easily. But just as the guides never arrange a trial beyond one's capacity to suffer it, neither do they place on one's table a lesson too difficult to be learned. Just keep trying . . . and ask your guides to help.

Libra

When Mars is conjunct Saturn in this cool but sociable sign, it is very likely that the individual will have to learn to interact more genuinely with others, especially those who are close by kinship or friendship. The person with this pattern usually has a tendency to think that others have deliberately excluded him, whereas in fact it is he himself who has closed the door. Expect some turbulence in marriage, and some loss or sorrow as well.

Scorpio

The Mars-Saturn connection in this sign usually points to one who must come to terms with his own sexual identity. The specifics cannot be given, since there are so many facets and conflicts in relation to the physical passions. Simply look for a life-pattern that has run aground on the shoals of sexuality, then determine how to use the Martian energy to make changes.

Sagittarius

When Mars conjoins Saturn in this sector of the Zodiac, the usual indication is that the individual tends to alternate between periods of youthfulness and enthusiasm on the one hand, and care-worn seriousness on the other. The task at hand is to become more consistent, and it is possible to use the Martian energy to accomplish it. The first step, of course, is to *own* the problem, to recognize that it is there.

Capricorn

In this worldly sign, both Mars and Saturn are well placed (even though each afflicts the other). Mars is exalted and Saturn is at home. As always, Mars in Capricorn points to a great capacity to work hard for success in the world's eyes. In the case of this conjunction, however, the energies available for any endeavor are somewhat restricted (the moreso when the conjunction is close, i.e. within 3 degrees). The task before the individual is to understand that the power and drive he has were not given to him to be totally funneled into the career. Some of it must be placed at the service of the emotional and affectional life, otherwise he will finish his days in loneliness and isolation.

Aquarius

When this conjunction occurs in Aquarius, the meaning is somewhat blurred: on the one hand it can point to the mental sphere as that in which to expect an alternation between energetic enthusiasm and exaggerated seriousness; on the other it can designate the emotional/affectional area, where the individual must learn to express love in a way that does not require a sexual component.

Pisces

The conjoining of Mars with Saturn in this watery, changeable sign identifies a person who must learn how to control his emotional outbursts. Until he or she can cap up this debilitating leakage of power, physical exhaustion or illness will continue to be in evidence.

Chapter 17

Mars-Uranus Conjunctions by Sign

At present, science recognizes the existence of three major planetary bodies beyond the orbit of Saturn, these being Uranus, Neptune and Pluto. None of these is visible to the naked eye. Although there do exist additional bodies beyond Pluto in gravitational orbit around the sun, the guides of individual human beings do not utilize them when planning a birth chart, simply because astrologers have no way of determining their positions. For additional commentary on the question of how guides "arrange" individual charts, see Chapter 8.

In this and the following two chapters, we shall deal with the significances of the conjunctions of Mars with the three trans-Saturnian planets.

Uranus

Uranus takes some 84 years to complete one circuit of the zodiac, spending about 7 years in each of the twelve signs. Before discussing the conjunctions, it is an interesting aside to contemplate the punctuality with which this planet has marked the major stages in the history of the television industry. Uranus is said to "rule" all high-tech developments, and as it has progressed from one sign to the next, so the nature of television has kept step in a fascinating way.

From about 1949 to about 1956, Uranus was transiting through Cancer, the sign ruling the home and family. Television began to become available in Canada and the United States around the beginning of this transit, i.e., it began to be introduced to the *home*. This was the early, experimental stage of its development, without any clearly defined themes. Then, from 1956 to about 1962, Uranus passed through Leo, traditionally connected with the idea of the stage, acting, and especially the "big star". That transit saw the emergence of a number of strong personalities with their own television shows, right in line with the Leo rulership. Following that, from 1962 until about 1969, Uranus transited Virgo, a sign relating to the mental area and ruled by the communication-oriented Mercury. This passage saw television begin to stretch its wings, opening up its reportorial possibilities. The most prominant development was the nightly reporting of the Viet Nam war.

The next phase was the 1969 introduction of color to what had been a black-and-white medium. This coincided perfectly with the movement of Uranus out of Virgo and into Libra—a sign ruled by the aesthetically oriented planet Venus. Libra governs all artistic endeavor, and the medium of television obviously became more aesthetically pleasing with this change.

Uranus spent a little under seven years in Libra, then entered the Mars-ruled sign of Scorpio in 1975. Scorpio, more than any other sign, governs sexuality, violence and bloodshed, and the logical thing happened: censorship was relaxed and programs began to show more and more of these themes to viewers.

From about 1982 until 1988, Uranus was in the Jupiter-ruled sign of Sagittarius. Jupiter governs expansion, multiplication and bigness, and during this time-span television literally expanded to circle the globe: a large number of T.V. satellites were placed in synchronous orbit around the earth, thus greatly increasing the number of programs available to individuals, networks and cable companies.

It can be seen from the foregoing brief capsule history that at least some developments taking place on the earth do indeed follow the general significance of planetary placements. In the following chapter we shall discuss the way in which the *film* industry has marched in lock-step with the sign placements of its traditional ruler, *Neptune*. First, however, we will describe the significance of the Mars-Uranus conjunctions in the various signs.

Aries

When Mars conjoins Uranus in this Martian sign, the meaning relates to unorthodoxy. The urge to buck authority is strong, as is the desire to live an unfettered life. The lesson is primarily to realize that one's freedom is not genuine if it requires others to be in chains. Seek to liberate everyone, not just the self.

Taurus

The conjunction in this sign points to a tendency to become involved in unusual relationships, often with an unorthodox approach to sexuality.

Gemini

Here, the Mars-Uranus conjunction designates the mental arena as that in which the unconventional energies will seek to express themselves. Look for a powerful mind and a need to control mental spats. Especially, expect a person who "dreams up" the most amazing ideas or plans. A strong innovative streak.

Cancer

In this watery, emotional sign, neither of these planets is well placed. The individual is likely to show some negative effects from the unorthodox and possibly combative energies he experienced in the early home.

Leo

The combination of Mars with Uranus in this proud and sometimes pompous sector of the zodiac marks one who will seek to "lord it over" others (at least until the negative effects of such behavior teach him to back off a little).

Virgo

When this conjunction occurs in Virgo, the primary lesson to be learned is that of controlling the urge to criticise others. This person may also harbor a secret dream of one day finding the absolutely perfect mate. Unfortunately, Snow White and Prince Charming don't live on this planet.

Libra

The conjoining of Mars with Uranus in this sign indicates that the area of partnership and marriage will be marked by strife and sudden, unexpected changes. Also, others may look on this individual as different or odd in some way, and may tend to exclude him for this reason.

Scorpio

In this Mars-ruled sign, the conjunction in question tends to point to sexuality as an area where the unusual or unorthodox energies will make themselves felt. Often this includes some unpleasant or traumatic event in childhood.

Sagittarius

When the conjunction appears in this sign, the indication is that the individual will feel a strong pull toward adventure, freedom, and the open road. Typically for those with this pattern, the early years are marked by a lot of moving around and changes of address.

Capricorn

Mars is well placed in this sign, and when it conjoins Uranus as well, there is a likelihood that the individual will make a mark for himself in an unusual or highly technical sphere of activity.

Aquarius

The conjunction under consideration, when falling in Aquarius, denotes a powerful mentality, and the ability to grasp subtleties in a flash. A preoccupation with sensual fantasies should be tamed.

Pisces

Here the conjunction is badly placed. The energy of Mars tends to be sapped by this watery, emotional sign, whose vibrations are also out of sync' with those of Uranus. The result is often a kind of exaggerated melodramatic trait, in which the power of Mars is directed into displays of emotion or feeling. If indulged in too much, this can lead to physical weakness.

Chapter 18

Mars-Neptune Conjunctions by Sign

Prior to discussing the significance of Mars-Neptune conjunctions by sign, it is instructive to consider the way in which the sojourns of Neptune in the various zodiacal sectors has coincided neatly with the development of the *film industry*.

The first point to be made is that Neptune represents all forms of illusion, i.e. things not being what they seem. This relationship derives from the connection of Neptune with the sea. Under water, all things seem distorted, murky, unreal. Even when looking down through the surface of water from a point above it, the bottom and all underwater objects appear closer than they really are.

What facet of current civilization contains more in the way of illusion than the film industry? Consider these facts:

Actors and actresses make their living by pretending to be people they are not (i.e. the film characters).

The backdrops and scenes against which the action takes place are usually props and screens that merely *appear* to be real.

Film action and sounds appear to the film-goer to be happening in real time, whereas they result from an electro-mechanical process involving a moving film of plastic.

The plastic film itself is made from materials derived from the petroleum industry, which is traditionally placed under the rulership of Neptune.

The process by which the screen picture appears to move is an optical illusion, based on the physiological characteristics of the human eye. The individual images are flashed one after the other so quickly that the eye does not perceive the separate frames.

Finally, many famous actors and actresses lead lives of total fantasy, living in houses and clothed in apparel that the average person could never attain.

The entire industry, one must agree, is a giant game of "let's pretend".

Consider now the various phases of the film business, and how they have been heralded by the sign tenancy of the planet Neptune. The astute reader will observe something of a parallel with the Uranus connection to television.

Neptune was in Cancer from about 1901 to about 1915. Cancer rules the "home", and this period of time saw many of the early developments of the film industry. These developments were mostly the result of independent inventors working in their *homes*.

From 1915 to 1929, the god of the sea was in the sign Leo. Astrologers will know that Leo rules the stage, play-acting, and the "superstar" concept. All films during this initial commercializing phase were silent movies. Lacking sound, these productions had to rely on overacting, sweeping gestures and animated eyebrows to convey the emotions being portrayed. And of course, this period saw the rise of many famous actors and actresses.

Then, with impeccable timing, Neptune in 1929 slid into Virgo just as the first "talkies" were being shown to unbelieving audiences. Virgo is ruled by Mercury, the planet associated with speaking, communication and the voice.

Neptune remained in Virgo until about 1944, when it entered Libra. This sign is ruled by Venus, the planet of beauty and aesthetics. Right on cue, films began to appear in color during the ensuing few years, clearly adding to the pleasure experienced when viewing a film.

In about 1957, Neptune vacated Libra in favor of Scorpio, a sign traditionally linked with sexuality and violence. The following 14 years saw more and more films exploiting both of these concepts.

The foregoing should provide sufficient evidence to support the connection of the film industry with Neptune. We return now to the task at hand: the interpretation of the Mars-Neptune conjunctions in the various zodiac signs.

Aries

In this sign of self-absorption and unruly energy, Neptune is badly placed. The conjunction with Mars will draw out the negative side of Neptune even more. This will tend to result in moodiness, depression and withdrawal from others. Of course, such uncongenial traits can be overcome by putting forth the required effort. Alas, all too often the individual with this conjunction placement will not have the determination or courage to change what he sees as an inescapable curse. "That's just the way I am," he might say, "I can't help it".

Taurus

This portion of the zodiac is associated with stolidity, stubbornness and emotionality. When Neptune conjoins Mars in this Venus-ruled sign, there is usually some deeply buried wish to be free of all responsibilities. This desire may never be voiced or acknowledged, but the individual will show by his actions and attitudes that it is indeed there. Often the yearning to be unburdened will surface in terms of the pastimes most enjoyed. For example, the inveterate camper forever planning his next junket could well have this conjunction. (Editor's note: this is evidently given merely as an example of a trait that could be found in one with this conjunction. At the present time, hardly anyone alive has this pattern in his chart, because Neptune has not been in Taurus since 1889.)

Gemini

Mercury rules this mentally-oriented and often unreliable sign. With Neptune here, afflicted by the Martian conjunction, one can expect some lack of clarity on the intellectual plane. Often one finds some deeply rooted misconception or conviction, which leads this person into some unusual activity that does little good for him or for others.

Cancer

Neptune is not badly placed in this watery and emotional sign, however the conjunction with Mars in the Fall of the latter will tend to draw out some unclear and erroneous assumptions in regard to the home and how a home should work. Such a delusion will inevitably involve one or both of the parents, and this individual may find his own later home somewhat uncongenial due to the misapprehensions which he amassed when he was a child.

Leo

The conjunction of Mars with Neptune in this sun-ruled sign often leads the individual to believe that he is someone "special", or that he has some momentous task to perform in life. He may distance himself from family and friends under the prompting of this placement, erroneously thinking that they could interfere with his destiny.

Virgo

Neither of the planets under discussion is particularly tickled to be in this sign, especially when conjoined to the other. As a result, one can expect the negative side of both bodies to be in evidence: a critical nature, a cutting tongue, moodiness and suspicion—any or all of these may be hampering the personal life of the individual. Of course, there is always the possibility of completely overcoming these characteristics, and with effort they can all be left behind.

Libra

Mars is debilitated in this sign, and the proximity of Neptune does little to improve the picture. Mars in Libra, even by itself, points to recurring episodes when the individual feels like an outcast, as if others were deliberately (though subtly) shunning him. When this pattern is further complicated by Neptune, his ability to perceive the truth about his relationships to others is impaired still further.

Scorpio

The red planet is one of the rulers of this powerful, manipulative and brooding sign. Being well placed, one can expect Mars to lend physical strength and vitality to the one with this placement. However, Neptune is clearly unhappy here, primarily because of the closeness of Mars (an affliction). As a result, experiences involving money and sexuality will be marked by self-delusion and misperceptions. This individual would do well to work on these aspects of his persona, for in this incarnation he has the inner strength and grit to accomplish a great deal in terms of self-regeneration.

Sagittarius

This sign is a kind of resting place for many planets, due to its rulership by the beneficent Jupiter. Both Mars and Neptune find this sector rather agreeable, to the extent that no serious inharmonies are experienced. The most one can say about the placement is that it might exaggerate the desire for freedom and the open road, due to certain minor misapprehensions which this individual has.

Capricorn

When Mars joins Neptune in this Saturn-ruled, punctual and work-oriented sector of the zodiac, there arises an imbalance: Mars is exalted in this sign, whereas Neptune is miles out of its element. Hence one looks for traits that can be traced to an afflicted Neptune. These are not hard to locate: unclear perceptions in certain departments of the life, moodiness and depression. The particular areas of clouded thought will depend on the house in which the conjunction is found.

Aquarius

The conjoining of these two planets in Aquarius usually leads to a vivid imagination and a colourful way of speaking. Here is the "character" par excellence, the unusual personality that makes a strong impact on everyone around it. The positioning of Mars usually gives the mental faculty lots of power, while the proximity of Neptune seeks to enervate the thought process and seduce it into wasting energy on daydreams.

Pisces

When Mars is close to Neptune in this watery, empathetic, emotional and self-deluding sign, there tends to be a lack of determination and grit. One usually finds a certain feckless languor, an indifference to life in general.

Chapter 19

Mars-Pluto Conjunctions by Sign

It is important to understand that many of the interpretations that follow are for conjunctions that will not be found in the birth chart of anyone now living. Pluto is so far away from the sun that it requires almost 250 years to make one round trip through the zodiac. At the time of this dictation Pluto is in Scorpio, having traversed only five signs since the beginning of this century (when it was in Gemini). Its orbit is unusually configured, with the result that this planet moves more quickly through some signs than through others. For example, Pluto requires only about 13 years to traverse Scorpio, but needs roughly 27 years to go through Taurus (at the opposite point of the zodiac).

We have some reason to believe that the present text will be made available throughout at least one full circuit of Pluto, and therefore we wish to comment on all twelve possible locations for the conjunction of Mars with this tiny, cold and distant planet.

Before proceeding, however, we wish to ensure that the reader understands the essential nature of Pluto, in astrological terms. Pluto is like an implacable force that sweeps all before it. It is not possible to simply stand pat, i.e. to remain unchanged, when Pluto is active in one's chart. If the learning and the changing are refused, then the individual will be delivered over to the coercive aspect of Pluto, wherein events and circumstances are encountered which literally compell the individual to make the step of growth that is expected. The more one resists, the greater will be the pain of Pluto's activity.

Aries

When the red planet conjoins Pluto in this fiery, impetuous and headstrong sign, the implication is that a great deal of very powerful energy is available. This gift of energy can be taken as a kind of test, to determine whether the individual with this conjunction can learn to use it in a positive and constructive way. By judging the severity of the afflictions to this conjunction, the Astrologer can determine to what extent the same gift of energy was misused in a previous incarnation. If very badly afflicted (for example by the square or opposition from Saturn), then the test is extremely important. If the lesson is not learned by the time of the first Saturn return (around age 28-30), then there is a strong likelihood that the power of the body itself will be taken away—through illness or some other means.

Taurus

In this Venus-ruled, fixed, earthy sign, the combination of Mars and Pluto can be taken as a pointer to a particularly intense love-nature. The word 'compulsive' would not be amiss in describing this individual. However, it is important in this life to balance passion with *com*passion, that is, to learn that the feelings and needs of one's partner are as important as one's own. Those who have this pattern in the birth chart will recognize immediately the trait to which we are referring, and no further elucidation is required here.

Gemini

In the sign of the heavenly twins, the conjunction of Mars and Pluto signals one characteristic above all others: the tendency to *think* too much about matters which are more the province of the feelings and affections. The mind is powerful, active and shrewd, and it will tend to dominate all activities unless the individual exerts great effort toward developing the feeling side of his nature.

Cancer

Mars is badly placed in this watery, emotional, self-absorbed sign. Conjoining Pluto, the red planet will signal a difficult early home life—one marked by storm, suppression and coercion. This unfortunate beginning is usually related to karmic patterns brought forward from other incarnations, although in some instances the experience is volunteered in order that the individual's parents or other relatives can learn some important lessons. The best advice that can be offered to one with this placement is to try to rise above the effect of the early trials to which he was subjected, and to forgive those who were responsible for the pain. Unless a true attitude of complete forgiveness and absolution can be attained, the karmic ties cannot be dissolved and at least some future lives will be marked by a similar adversity.

Leo

Pluto is not badly placed in this proud, fiery sign, however the conjunction with Mars (the Lesser Malefic) will certainly seek to bring out some of its darker shadings. The primary trait to watch for is that of excessive *hauteur,* an overarching pomposity and arrogance that will intensely irritate many of those with whom this individual comes in contact. This characteristic may lurk beneath the surface in the early part of the life, but given the right circumstances it can burst forth at a later point, bringing in its train all that can be imagined of strife, disagreements and estrangement from others.

Virgo

This is a problematic placement for both Mars and Pluto. When the two planets are conjoined in this intellectual but practical sign, watch for a strong critical trait, and a soaring idealism which measures all others against an impossible archetype of perfection. Sadly, he will discover early that all humans are flawed, and this individual will end up deeply disappointed with life unless he can learn to cherish the quirks and foibles of others, and to love them despite their faults.

Libra

This is a sign that has an instinctive feeling for the "rightness" of the pair-bond. When Mars and Pluto are found together in this Zodiacal sector, the meaning is one which touches deeply the natural desire to have a mate. Mars will tend to bring about a storminess in the partnership department, while Pluto will add a coloring of obsessiveness and coercion. Here is a maxim to be pondered by anyone who has this conjunction: If one wants a bird to perch on one's hand, the worst thing one can do is to grasp and hold it.

Scorpio

Scorpio is jointly ruled by Mars and Pluto, both of which are powerfully placed in this brooding, emotional sign. Expect the main Scorpio traits to be to the forefront with this pattern, particularly the deep desire to control others and the environment. Scorpio is not happy unless it holds the reins tightly in its hands. Being at the mercy or whim of another is distressing in the extreme.

Sagittarius

The meeting of Mars and Pluto in this carefree and eternally young sector of the zodiac marks an individual who must learn to live in a world that does not measure up to the ideals which he cherishes in his heart. Pluto always points to some lesson that must be learned or some change that must be experienced, and in this placement the thing to be learned is to *accept* the world as it is. If the lesson of Pluto is not voluntarily learned, then the life-pattern will *compell* the learning.

Capricorn

Mars is exalted in this worldly, energetic and enduring sign, however Pluto is uncomfortable here. To this discomfort is added the affliction from the Mars conjunction. The result is a life marked by repeated episodes in which the individual feels *coerced,* either by others or by circumstances. Often the coercion will be noted in the department of life indicated by the house position of this conjunction. Because of the Capricorn connection, most such instances will also have something to do with authority figures such as parents, the police, government, etc. The main lesson designated is that of learning how to cope with the "musts" of life.

Aquarius

In this clear-minded but emotionally restrained portion of the great circle, the conjunction of Mars with Pluto will point to one essential trait: namely that of preferring mental exercise far ahead of physical or emotional experiences. There may be a tendency to view these other realms of activity as frought with danger, especially the danger of muddying the clear intellectual waters in which this individual loves to play. Unfortunately, Pluto is again pointing to a lesson that *must* be learned in this lifetime. The lesson is that mental gymnastics alone cannot lead one onto the high plateau of spiritual achievement, and that intelligence must be tempered and balanced by compassion and love before one is ready to leave behind the repeating cycle of life and death as it is played out upon this earth.

Pisces

Mars is badly placed in this watery, emotional and often pessimistic sign. Pluto understands the makeup of Pisces, but is unhappy with the tendency, often found there, to give up on life. When these two planets are together in Pisces, the main lesson to be learned is not to allow surrounding circumstances or the words of others to influence one's opinions, actions and/or feelings. Life should spring from *within* the individual, and not from outside, for the wellspring of all vitality, thought and emotion is to be found only within the crucible of the *self*.

Chapter 20

Mars Conjunctions by House

In this chapter we will give the general departments of life to which the various house placements of the Mars conjunctions pertain. The skilled Astrologer will be able to blend this information with the interpretations given in previous chapters regarding the underlying meaning of the various Martian contacts.

The houses in a birth chart are determined with reference to the Midheaven and what is called the Ascendant. The latter is defined as the skypoint rising on the eastern horizon, and is taken as the "start" of the house systems that are the most widely used in western Astrology. Planets located at or near the Ascendant location are to be given more weight than planets not so located. The opposite point, i.e. the western horizon, also highlights any planets located close to or on it. There exist two additional points of importance in a typical birth chart, namely the Midheaven and the Nadir. The Midheaven is that location on the zodiac which lies due south at the time in question (usually the time of the "first breath"). The Nadir is the point opposite the Midheaven.

Because the axis of the earth is not exactly at right angles to the orbit of the earth around the sun (but rather is angled about 22 degrees from a true right angle), the zodiac location of the Midheaven does not remain constantly 90 degrees away from the zodiac position of the Ascendant. Instead, at least for latitudes which are some distance from the equator, the Midheaven oscillates, so-to-speak, from one side to the other of the actual zodiac point which is 90 degrees from the Ascendant. Thus, each of the four quadrants in a chart will smoothly fluctuate, through any day-long period, between a span that is less than 90 degrees and a span that is greater than 90 degrees.

Over the centuries, Astrologers have evolved several different systems by which to carve up the different quadrants into three houses each, thus producing twelve houses in total. The simplest method is to take each house as representing a span of exactly 30 degrees. This is called the "Equal House" system, and has the disadvantage that the 10th house does not necessarily start at the location of the Midheaven. Another house system is that of Placidus, in which the divisions are done on the basis of time intervals. There are several other house systems, but except for one of them, they do not require mention in this book. The one to which we refer is that of Porphyry, because that is the system which humanity's guides use in planning a birth chart for a particular soul.

We are fully aware that all Astrologers have their pet systems of house division, and that many practitioners are likely to regard the statement at the end of the preceding paragraph as being rather arbitrary. We certainly understand that attitude, however our mandate in this book is to tell the truth as we see it, and that is what we are doing in singling out the Porphyry house system as the preferred one. We think that any Astrologer will be in a position to test our statement for himself, and provided he has the proper hour and minute of the first breath, we believe that he will be persuaded by the sheer workability of the system.

It is important to understand, in connection with the Porphyry system, that the beginning point or "cusp" of each house is taken to be the most prominent part in terms of highlighting planets in the chart. The way to view the matter is to consider each house to be one of twelve spokes in a wheel. Each spoke is a house cusp. Any planet near one of the spokes, whether in the actual house or the one next to it, has an interpretation based on the house of which the spoke is the cusp. This is becoming complex, and we think an example will illustrate what we mean. Suppose that the second house of a chart has its cusp at 13 Gemini and that the cusp of the third house is 7 Cancer. This means that the span covered by the second house is 24 degrees. Now a planet at 13 Gemini will be exactly on the cusp of the second house, and will therefore be the planet with the strongest second-house meaning. Suppose this planet to be Mars. Suppose further that Mercury is located at 11 Gemini and Saturn is at 16 Gemini. This will place Mercury in the *first* house, but very close to the end of the first, in fact only 2 degrees away from the beginning of the second. Saturn will be in the second house, only 3 degrees away from the cusp. In our approach to interpreting the Porphyry system, we would consider Mercury to have primarily a *second* house influence, even though it is physically in the first house. This is due to its closeness to the cusp of the second. If Mercury had been at 7 Gemini, we would view it as a first house planet, and too far away (6 degrees) from the second to have any second house influence. Planets *within* the second house, up to about 3 or 4 degrees from the end of the house (7 Cancer), have second house meanings in descending order of importance as they appear further away from its cusp. Thus, in our example, Mars has the strongest second house influence, while Saturn and Mercury have about equal significance for the second house. If the moon, say, were at 25 Gemini, it would be further from the cusp than Saturn, but still clearly within the second house. Accordingly, it would have a lesser influence or significance for second house matters than any of Mars, Mercury or Saturn.

Indeed, we have just painted a portion of an intriguing chart, which we will attempt to interpret for the benefit of students and practitioners of Astrology. The second house, as explained below, primarily governs the following departments of the native's experience ("native" = the person whose chart it is): the affectional life, money and possessions. Many texts ignore the affections when defining the rulership of the second house, and we wish to correct this omission here.

Now, we have stipulated a chart in which the following planets have a second house coloring, in descending order: Mars, Saturn, Mercury (even though in the first house), and the moon. Looking first at the affectional life, the placement of Mars precisely at the second cusp (13 Gemini) will certainly produce a strong sex drive and a certain storminess in this area. The conjunction of Mars and Saturn is in the mentally oriented sign of Gemini, and by blending the interpretation given in Chapter 16 of this book with the second house rulership of the affections, we can assume that at least the storminess and strife in this area of experience will arise due to the native's tendency to verbally abuse or attack others (Mars afflicted in Gemini by the conjunction of Saturn), and/or to the native's inclination to become depressed or overly serious about things (Saturn in the mental sign of Gemini, afflicted by the conjunction of Mars).

With Mercury also in the picture, forming a triple conjunction with the two Malefics (Mars and Saturn), the "Gemini connection" is reinforced—i.e. due to the fact that Mercury rules this sign. The student may wish to review our earlier book, *Astrology Plus,* for interpretations of these three planets when in Gemini. The heavy aspects to Mercury will certainly tend to promote the Gemini shortcomings of scattered forces, unreliability and verbal abusiveness, and any or all of these may become a factor in the affectional problems that can be expected.

The moon at 25 Gemini is very wide of Saturn, and we would not consider the pattern to constitute a conjunction. Regardless of aspects, the moon produces *changeability* in any life department ruled by its house placement. Hence, we would look for much fluctuation in the love-experience and also in the financial fortunes. Again, we suggest that the student review the interpretations for the moon in Gemini and in the second house that are found in our earlier book, *Astrology Plus*.

That completes our excursion into chart interpretation, save for one final revelation which we have permission to impart. This has to do with what may be called "The Rule of Tripling".

The Rule of Tripling

In any given birth chart, each planet is located at a particular point on the 360 degree zodiac. These points are defined in terms of the degrees and minutes of the various signs, each spanning 30 degrees. Thus, we might have Mars at 7 degrees 20 minutes of the sign Cancer in the first house, and the moon at 2 degrees 30 minutes of Capricorn in the seventh house.

In this scheme, the specific degree/minute location is multiplied by 3 to arrive at a specific *age* at which one can expect some circumstance or event which is foreshadowed by the planet concerned, taking into consideration its aspects, its sign and its house.

Using the example given above, we have Mars at 7 degrees 20 minutes of Cancer in the first house. The only aspect we know about is the opposition from the moon in Capricorn. This constitutes an affliction to Mars. Multiplying the whole degrees by a factor of 3 gives 21 ($3 \times 7 = 21$), and multiplying the minutes by 3 gives 60 minutes, or one degree. Thus the age designated by the tripling rule is exactly 22 years. To ensure that this principle is understood, let us explain the matter another way: The position of Mars is 7 degrees 20 minutes. That is exactly 7⅓ degrees. Multiplying 7⅓ by 3 gives exactly 22. This will designate a time close to the 22nd birthday, during which the life will be touched by some event or circumstance foreshadowed by Mars. Now, with Mars in the first house, one can be reasonably sure that the ego or self-picture will be affected in some way. The sign Cancer points broadly to the early home experience. Since Mars is related to strife and conflict, as well as sexuality, and with the opposition from the moon in the seventh house (marriage, partnership), this pattern paints the picture of some trauma around the 22nd birthday, a trauma which may combine any or all of the areas just defined: the early home, the ego or self-picture, the marriage, sexuality, partnership, and one or both parents (from the moon in Capricorn).

The same exercise can be gone through with the moon, at 2 degrees 30 minutes of Capricorn. The triple rule applied to this placement produces an age of 7½, when some event or circumstance can be expected that will be identified by a combination of some or all of the key ideas suggested by the moon afflicted in Capricorn in the seventh house. We will leave the student to run through this exercise by himself.

We come now to our last duty in this book, namely to set forth the main house rulerships as we see them from this vantage point. For the most part, the information to be given below duplicates the current understanding of students and practitioners of Astrology.

First House

The first house in the birth chart, that sector extending counterclockwise from the Ascendant, designates essentially the self-picture or ego of the individual. The sign on its cusp and the planets it contains can be taken as pointers to certain facets of the way the person *sees* himself or herself. The secret of analysis for the first house is to consider the *general* meaning of a planet or conjunction in a given sign, and then look for ways in which that general meaning can manifest in connection with the ego or self-image. Take the following example. Suppose an individual has Mars conjunct Uranus in Aries and in the first house, relatively close to the Ascendant. The trait broadly signified by this conjunction in Aries (see chapter 17) is that of a strong drive to live an unfettered life, a leaning toward unorthodoxy, and an inclination to buck authority. That is the meaning regardless of house. However in the *first* house, this meaning narrows down: these traits are those of the *image* which the person has of himself. Now, that image may or may not reflect reality at a higher level; in other words, the soul itself may not regularly project these drives. Nonetheless they are present *now,* and as such they will strongly affect the course of the life-pattern. Because the defense of the self-picture is of paramount importance to every soul, this individual is likely to be very insistant upon his right to live the way he chooses.

A further point should be made here: the tendency to see himself as unorthodox might be something that was felt to be *needed* for this soul. In previous incarnations, there could have been a tendency to be fearful, conventional and retiring—so much so that the soul was being hampered in its desire to improve and broaden its horizons of experience. If this was the case, then the guides of the individual might well have suggested that the Mars-Uranus "overlay" effect (caused by arranging birth when the conjunction of these two planets in Aries was about to rise over the eastern horizon) could give to the individual the needed pluck and courage to strike out on new and unusual paths. We believe the foregoing examples will suggest the best way to approach the matter of house interpretation, and accordingly we will now run through the basic meanings of the remaining houses, as we see them. Where appropriate, we will continue to illustrate the explanations with the Mars-Uranus conjunction in Aries. Bear in mind that these correlations work best with the Porphyry system of house division.

Second House

The second house relates to the affectional nature and love-experience of the individual, and only secondarily to his money and possessions. It points specifically to the *emotional* or *feeling* aspect of the approach taken to the "mate" or partner in the affectional area. This is to be distinguished from the three other basic aspects of affection, which fall under the fifth, eighth and eleventh houses. Should the Mars-Uranus-Aries conjunction fall into this house, the suggestion is that the unorthodoxy and the yearning to be unchained will to some extent interfere with the smooth running of the affectional life. There may be a strong disinclination to enter into a long-term relationship, because a tie of this kind is viewed as suffocating or limiting.

Third House

This is essentially a mental house, in the sense that it designates aspects and traits of the mind of the individual. The presence of the Mars-Uranus-Aries conjunction in this house would suggest that the unorthodoxy touches mainly the way the individual *thinks*. For example, he/she may hold unusual philosophical beliefs, or may simply be a person who reasons in a unique or uncommon way.

Fourth House

This house is extremely important—at times even more important than the first house. Students and practitioners of Astrology will know that both the first and the fourth belong to the set of four houses called "angular". The other two are the tenth and the seventh. Being angular, of course, all four enjoy prominence in a chart, however we wish to pass on something new in regard to the fourth house, which will cause many Astrologers to view it in a different light. Specifically, the fourth house is to be regarded as designating the manner in which the individual sees not only his parents, but *himself* as an offshoot or "product" of his parents. It is important to understand that there are many factors which go to make up the basic package of the "self" which is living the incarnation. The first house designates certain of the traits that contribute to this complex of the self, but it must be remembered that these first house characteristics are not necessarily a strong part of the ultimate soul. As explained in the example given earlier for the first house, the planets or conjunctions noted in the first house may be placed there deliberately in order to supply some antedote, some corrective factor to balance a tendency which is regarded as counter-productive. In other words, the first house interpretation may only be an overlay or filter which affects the individual only for the present life.

But there are other factors which go into the package of the self. One of these is related to the traits picked up from the interaction with the parents (or simple observation of them), usually in the early home. In our book, *Childlight,* we explain what can be called the role-switch reaction (page 34 of *Childlight*—Ed.). This phenomenon arises when the child is punished severely in a manner that the child feels is not warranted by his transgression. The resentment of the punishing parent typically leads to the following thought in the youngster: "I wish I was big and you were small, and then I would hit (spank) you hard just like this." Repeated episodes of overreaction and overly severe punishment on the part of the parent will reinforce that thought-complex in the mind of the child, and cause it to become deeply rooted in the subconscious . . . until the day many years later when the child, now fully grown, is presented with a scenario that matches the thought. His own child is caught doing something that triggers the thought-form in the parent's mind, and as a result the little one is punished just as if the parent were "hitting" back at his own parent for that long-ago beating. This is a particularly prominent example of the effect of one's parents on one's self or self-picture. However, there are many other less noticeable ways in which each individual on the earth plane is influenced by the parents. And at least some of these are typically indicated by the nature of and aspects to planets in the fourth house.

Fifth House

Here we come to another of the four "affection" houses. The fifth speaks of the affectional relationship with a) parents and b) children. The reader will readily recognize that the love-connection with parents and children is quite different from the more intimate kind of affection that is felt for the mate or partner. Many Astrologers attribute "love affairs" to the fifth, but in an important sense this correspondence is erroneous. When a soul has *ignored* the relationships with children and parents in favor of the "love-affair" type of involvement, then it may be appropriate to place afflicted planets in the fifth. These would be meant to show the manner in which the "lesson" of the fifth house will be taught (the lesson being not to downgrade or ignore the importance of feeling affection for children and parents). To use our example from previous entries, suppose that Mars and Uranus were found in the fifth house, in Aries. The afflicted Uranus (afflicted by the conjunction with Mars) would incontestably point to the likelihood that love-involvements with partners would be unusual and subject to sudden changes and unexpected endings. The position of Mars in the fifth, with or without the Uranus conjunction, would indicate a problem in dealing with a parent, a child, or both. It would also inject some coloring of conflict, rivalry or ill-will into the love-liaisons.

Sixth House

The sixth sector of the house pattern points to the individual's ability to deal with routine chores and practical details. Any planets in this house can be looked on as modifying or explaining some facet of the individual's abilities in this area. Let us offer again the example of Mars-Uranus conjoined in Aries:

We have pointed out in *Astrology Plus* (available from Marcus Books—Ed.) that the sixth house is basically a positive placement for the planet Mars. In this sign, the red planet tends to sharpen and clarify the mind, while at the same time improving the organizational abilities. Uranus is strongly afflicted in this position, however, due to the proximity (conjunction) to Mars. An afflicted Uranus tends to work *against* the organizational abilities. Thus we have two contrary indicators within this placement. The Astrologer would have to weigh the relative strength of Mars and Uranus in a case of this kind, judging which one is more strongly positioned—by the number, beneficence and closeness of other aspects to these two indicators, and by noting which of the two planets is closest to the Porphyry cusp of the sixth house.

Seventh House

The seventh house points essentially to "the other": specifically the mate or love-partner. When afflicted planets are found in this portion of the natal chart, one may be sure that there will be stresses and disruptions in this area of life. Using our example of Mars-Uranus-Aries, we would look for storminess in the marriage/partnership area due to Mars, and sudden breaks and surprises due to the afflicted Uranus. The apparent cause of these difficulties would typically be found in the unorthodox approach of the individual with this placement, and his/her strong inclination to live an unfettered life.

There is something further in connection with the seventh house. Just as with the fourth, the seventh does modify or contribute in some manner to the self-image. Specifically, those traits that are acquired by emulating or copying *others* will tend to be foreshadowed by the identity and aspects of any planets in the seventh sector. Thus, using our now well-worn example of Mars/Uranus/Aries, we should expect the individual to reflect an acquired trait of combativeness or aggression due to the position of Mars in the seventh, and a desire to be unorthodox and unchained due to Uranus. These characteristics would be at least partly the result of an interaction with some other person(s), during which they were "picked up" by the individual whose chart we are doing. In most cases, the traits would be acquired from peers or siblings, since characteristics deriving mainly from the parents are shown in the fourth house.

Eighth House

At the present time in humanity's journey, the eighth house is associated with sexuality and "the money of others". However, the true significance of this sector of the zodiac runs much deeper. This house is related to that kind of ultimate regeneration which comes through death, or through a "dark night of the soul", or through honest and conscious effort toward casting out imperfections.

Dealing first with the sexual connotation of the eighth house, the reader will realize that here is another of the four kinds of "love" available to humanity. Astrologers will understand that the second, fifth, eighth and eleventh form a cosmic cross in the sky, which is appropriate because of the association of the Cross of Calvary with God's own love ("For God so loved the world that He gave His only begotten son . . ."). We will summarize this correlation later in our remarks on the eleventh house.

We would now like to consider the meaning of our example conjunction (Mars and Uranus in Aries) in the eighth house. Mars of course will point to the sexual passions, when placed here. As we have explained in our book, *Astrology Plus,* this placement of Mars always requires the individual to go through periods of enforced abstinence, in order to prompt him to explore the other facets of human affection. Taken by itself, Uranus in this sector when afflicted (by the conjunction of Mars or in any other manner) suggests financial loss through the partner's actions. However, the particular conjunction under consideration—associated with unorthodox behavior and a desire to live an unfettered life—will, when found in the eighth house, manifest through a tendency to seek uncommon sensual experiences, along with a determination not to be enslaved by them.

Ninth House

This sector of the chart represents the modalities used to *broaden* the mental horizons. The "early" stage of such broadening comes about through travel experiences, in which the individual is presented with new information through contact with other languages, societies and customs. The "mature" phase of mental broadening arises through study and contemplation. When the example conjunction (Mars-Uranus-Aries) appears in the ninth, it strongly supports a likelihood that the individual will become involved in a great deal of moving about, with an active, inquiring mind. Generally, planets in the ninth indicate a strong moral fibre, and Mars here tends to point to the "crusader" approach to life. The "unorthodox" tendencies of the conjunction are likely to be found in the belief-system of this individual, while the freedom-loving trait will reinforce his love of the "open road".

Tenth House

Here is the last of the four houses which provide clues to deciphering the self-picture or ego of the individual. We have pointed out that the first house suggests both the innate qualities and those that arise through astrological overlays. The fourth house points to qualities derived from conscious or unconscious interaction with one or both parents. The seventh house indicates traits acquired from siblings and peers. In the tenth, however, what is shown are the characteristics which are handed down *genetically* from one's forebears. Of course, this house offers a great deal more (principally relating to the career or role in life), but these additional aspects of the tenth sector are dealt with in many other books on Astrology. We merely wanted to make certain that the genetic connection through the tenth was revealed. The reader may wish to review our discussion, in *Astrology Plus,* of the meaning of the sun in the tenth house.

Using our example of Mars-Uranus-Aries, we would expect an individual with this tenth house conjunction to have an unusual genetic background—one which may have contributed to his feistiness and self-assertive qualities. Furthermore this scrappiness, unless controlled, may lead to career upsets and losscs. Thc afflicted Uranus warns against "taking chances" in the career area, as sudden and unwelcome changes could result.

Eleventh House

As we point out in *Astrology Plus,* the eleventh house is largely misunderstood by Astrologers. It is labeled the house of "friends", whereas in fact it is the house of "former enemies". Planets in the eleventh always designate individuals who in past incarnations were adversaries. The hope is that the individual with eleventh house placements can learn to get along with such previous enemies. When the planets have essentially positive aspects, the task of learning affection will not be difficult. However, placements that are mainly afflicted point to individuals for whom love has not been felt for many lives. Now they have resurfaced, and it would be wise to put real effort into repairing the bridges of affection that have fallen into disrepair.

The eleventh is the last of the four "love" houses. We have pointed out that the second designates emotional love for the mate or partner, the fourth refers to the love felt for parents and children, and the eighth points to the physical or passionate aspect of love. To complete the pattern, the eleventh pertains to the *reverse* of love (hatred or dislike), and also the feeling of *friendship* or comradeship that can result from the transmutation which this house promises to those with the will to accomplish it.

Finally, we wish to discuss the placement of our example conjunction (Mars-Uranus-Aries) in the eleventh sector. We would expect the individual, at some time in his life, to encounter someone of a very unusual nature for whom an instinctive dislike or revulsion would be felt. The interesting thing is that the native (the person whose chart it is) would *himself* possess many of the same characteristics that repell him in this other person. What is happening is that the native's own unorthodoxy is being held up to him like a mirror, in the hope that he will acknowledge its existence.

Twelfth House

The first task before us is to show how all of the "cadent" houses in a chart pertain to mental aspects of the individual. The third house represents the gathering of information and data, the person who continually seeks new things to learn. The sixth has to do with the organization of that data into an orderly arrangement, with a view to understanding its true significance. The ninth house leaps beyond the simple gathering and organization of information, as it seeks to see and comprehend aspects of existence from a more philosophical or metaphysical point of view. Note, however, that the ninth house thinker remains closeted within himself, trusting to internal processes to reveal the truth to his mind. Finally, there is the twelfth house. And here we find available the only true access to ultimate truth.

The twelfth house mind knows how to reach out to sources *beyond* the confines of the individual conscious mind, to tap wellsprings of knowing that the everyday world does not accept. In its most positive aspect, the twelfth house points to intuitive sources, channeled information and direct revelation. However, in its negative manifestation, the twelfth can designate depression, melancholy and mental imbalance.

We have pointed out in previous writings that the twelfth house shows certain of the contents of the subconscious, as well as providing some information regarding the most recent previous life. Taking our example conjunction of Mars and Uranus in Aries, we would expect this pattern, if placed in the twelfth, to point to a previous, conflict-ridden, male (Mars) life that ended suddenly, unexpectedly (Uranus), and probably violently. If the planets are in the early degrees of Aries, then the life would have ended at an early age. In addition, this individual would certainly be expected to manifest the unorthodoxy and freedom-loving aspect given in an earlier chapter for this pattern. However with its placement in the twelfth, he would tend to keep these traits to himself. (As most Astrologers know, the twelfth also designates those areas that are private, secret or hidden in some way.)

This completes our survey of the houses in a birth chart, and provides a useful method of interpreting the house placement of various planetary indicators.

Epilogue

This book has afforded us the opportunity to pass along information additional to that which has already been transmitted in connection with Astrology. It has also provided a suitable forum in which early investigations by our channel could be summarized. As such we believe this work will arouse interest not only among those who have looked in some depth at the Science of the Stars, but among laymen as well.

In the months and years ahead, the consciousness of humanity is destined to grow and expand. There will arise events and circumstances which cannot be explained from within the narrow prejudices that now characterize most religious, social and political philosophies. The writings we have offered through this source are able to provide a vision of reality broad enough to encompass all of what mankind's future holds, and we are pleased that this series of books will continue to be made available.

As to the planet Mars, we see a day—perhaps not too far into the future—when the influence of this body will no longer be able to engender warfare and bloodlust. Humanity will have learned to transmute all destructive energies, building from them a new power to heal and bless this planet and the many lifestreams that call her their home.

To those who would have a part in the great drama of resurrection we say, be of good cheer. The tide of destruction must have its brief time, but then will recede before mankind's collective determination to build the world anew. Like the Phoenix rising whole from the ashes of its former self, the human spirit will soar skyward on wings of glory, and the evil wind that has brought the race to the brink of ruin will never blow again.

May the peace and blessing of all the higher beings who care for humanity's struggle be with you forevermore.

OM MANI PADME HUM

MAILING LIST

For a free catalogue of books, audio and visual tapes and other items, write to the address below. You will also receive future mailings regarding speaking tours by M.B. Cooke, workshops in the Hilarion Material, and our regular special offers of new books and tapes at reduced prices.

Marcus Books
P.O. Box 327
Queensville, Ontario
Canada L0G 1R0